Wise, Chen, and Yokely
  Microcomputers: A Technology Forecast and Assessment to the Year 2000

Krone
  Systems Analysis and Policy Sciences: Theory and Practice

Systems Analysis and
Policy Sciences

Systems Analysis and Policy Sciences

THEORY AND PRACTICE

Robert M. Krone

University of Southern California

A WILEY-INTERSCIENCE PUBLICATION

JOHN WILEY & SONS   New York • Chichester • Brisbane • Toronto

The photographic plate for the book jacket is a reproduction of *The New Horizon,* a 1978 acrylic painting by Homer E. "Heg" Gutierrez, Philippine artist. It was included in his second one-man exhibition, "Heg Perceives the Year 2000," held in Manila, Republic of the Philippines, September 1-15, 1978. Gutierrez's paintings are visual expression of his intimations of things to come in the year 2000. Heg Gutierrez identifies with the concern of the Spanish philosopher Ortega y Gasset that unless man reexamines and restructures his priorities in the face of his overdependence on science and technology, he will be reduced to the level of a machine and the machine will be elevated to the level of man. That concern is shared in this book through the assertion of systems analysis being a combination of art and science and the stated preference for political processes over scientism. The photograph of Gutierrez's painting is by Robert Krone.

Library of Congress Cataloging in Publication Data:

Krone, Robert M   1930–
　Systems analysis and policy sciences.

　(Wiley series on systems engineering and analysis)
　"A Wiley–Interscience publication."
　Bibliography: p.
　Includes index.
　1. Decision-making. 2. Management.
3. Policy sciences. 4. System analysis.
I. Title.
HD30.23.K76　　658.4'032　　80–13335
ISBN 0–471–05864–5

Printed in the United States of America

10 9 8 7 6 5 4 3 2 1

Contributors

Larry J. Campbell, Systems Analyst, RCA, Kwajalein Missile Range, Marshall Islands, M.S. Systems Management

Fortunato F. Cataldo, Chief of Performance Assurance, Kentron International, Inc., Kwajalein Missile Range, Marshall Islands

George Chen, Manager, Taiwan North-South Freeway Project, Republic of China

Caroline Chung, B.S. Commerce, Soochow University, Taipei, Republic of China

Homer E. "Heg" Gutierrez, Artist, Manila, Republic of the Philippines

Heinz Kesterman, Ph.D. General Manager of German Remedies, Taiwan Ltd., Republic of China

William M. Kinney, Lieutenant and pilot, U.S. Army, M.S. Systems Management

Kenneth Kou, Manager, Taiwan North-South Freeway Project, Republic of China

Arthur Sommers, Capatain, U.S. Army, B.S. in Business Administration

V. E. Spencer, M.S. Chemistry (Organic Chemistry), Soils Research Chemist, University of Nevada (Retired), Reno, Nevada

Clarence H. Walker, Jr., C.M., Chief of Plans and Programs, Production Engineering Control Department, Global Associates, Kwajalein Island, Missile Range, Marshall Islands, Master of Science, Systems Management

Hsi-i-Wang, Graduate student, USC M.S. in Systems Management Degree Program, 1977, Republic of China, B.S. in Civil Engineering

To Sue

and to Kat, Bob, and Don,

wending their way through a complex world

SYSTEMS ENGINEERING AND ANALYSIS SERIES

Since the start of the Wiley Systems Engineering and Analysis series in 1965, there have been many significant changes in our increasingly interdependent world that affect the technological systems to which the series was dedicated. Societal concerns with the natural environment and the damage being done to it, technical concerns for our increasing energy and materials needs and our finite energy and materials supply, financial and economic concerns with the problems of high inflation and continued high unemployment, social concerns for the opportunities for women and minority groups as well as for the third and developing worlds; these are representative of some of the broader systems problems that are becoming more and more important.

As editor of the Wiley Systems Series I was pleased to receive Dr. Robert Krone's manuscript. His emphasis is on political and societal issues. He has purposefully minimized the quantitative and mathematical emphasis in understanding and solving systems problems.

His use of case problems with international examples featuring the Pacific region brings new geographic areas to our attention. The use of solutions developed by students, women, and persons from developing countries broadens our perspective as to who can use systems analysis and where and how such analysis can be applied.

HAROLD CHESTNUT

October 9, 1979

Foreword

Any serious thinking on contemporary and emerging problems and issues leads to the conclusion that present assumptions, perspectives, orientations, institutions, knowledge, and methods are inadequate and increasingly obsolescent. Energy and food scarcities, erosion of governmental authority, ethnic conflicts, Crazy States, growing hiatus between haves and have-nots, new aggressive ideologies, proliferation of mass weapons, global multichannel communications, potential blessings and curses of genetic engineering, interaction between more than 150 nominally sovereign and independent states—these are just a few of the features that make the present quite different from the past and that promise to make the future even more different from the present. Historic discontinuities, trend mutations, catastrophic (in the technical sense) jumps, and accelerating ultrachange do make traditional modes of incremental policymaking and piecemeal social intervention at best inadequate and frequently counterproductive. Required instead are order-of-magnitude transformations in human and governmental capacities to appreciate situations and design innovative grandpolicies.

I think it is not exaggerated to speak about the need for a novel decision-culture; this is essential if humanity is to handle its increasingly difficult problems. This new decision-culture must be attuned to congenital characteristics of contemporary and future issues, especially to uncertainty and complexity and novelty. To face such features, there is need for radically improvement in our capacities to engage in fuzzy betting, to manage complexity, and to practice sociopolitical architecture.

Building up such capacities involves a broad variety of activities, ranging from educating young children in comprehending murky complexity to developing new modes of citizen participation in collective choice, from restructuring the international state system to seeking breakthroughs in artificial intelligence. Within the multidimensional sets of such required activities, and especially critical subset includes the advancement of policy knowledge and feeding it into a new type of statecraft. Novel knowledge based on revolutionary paradigms, innovative power structures, and newly designed bridges between power and knowledge—these are among the requisites for equipping humanity to try and manage its future, or, at least to prevent irreversible massive damage.

To move from desiderata to stocktaking, some promising developments can be discerned, though still in their initial phases. Among these, systems science and policy science in their various forms may provide a basis for producing some of the necessary knowledge, think tanks may serve as pillars for the bridges, while university programs in public policy provide precedents for training policy experts. But the distance between maximum contemporary achievements and minimum urgent needs is still much too large. Systems and policy sciences are primitive, think tanks are subdued, and public policy programs train technicians rather then educate professionals capable of formulating statecraft.

Overcoming this gap and progressing toward a new decision-culture involve a multiplicity of activities. One of these is the preparation of good texts in systems analysis for decisionmakers. This is the distinguished contribution of *Systems Analysis and Policy Sciences.*

Let me recommend that the reader and student of this text bear in mind the following points:

—Pay attention to the interplay between theory and practice. Constant iterative practice in moving between scientific ideas and the real world is an important way to advance useable policy knowledge.

—Bear in mind the interdependence between explicit knowledge and extrarational processes. Achievement of a synergism between these two is a must for statecraft, as well as for executive systems analysis.

—Consider carefuly the cultural variables. The tendency to see others as a mirror image of oneself is a major fallacy of Western policymaking, as illustrated by many intelligence failures.

—Move between the policy level and the metapolicy level. Improvement of the policymaking-system is both a condition for better discrete policy decisions and a consequence of them; the latter may be more important than the improvement of one or another single decision.

—The bibliographic notes provided in this book are a guide to essential readings, not merely a list to be glanced at. Anyone so presumptuous as to wish to engage in systems analysis in a complex world has a moral duty and professional obligation to engage in intense learning all his or her life.

One of the most successful of modern statesmen noted that even a very long march is composed of single steps. This book by Robert M. Krone is more than a single small step—it is a long step in the right direction.

YEHEZKEL DROR

The Hebrew University of Jerusalem
January 1980

Preface

This book presents an expanded view of Systems Analysis as an essential set of useful concepts and techniques for analysts, for managers, for politicians, and for civil and military decisionmakers in complex public or private human systems. It is also written for university professors teaching those persons at the graduate level. My interest is in applied science for the improvement or design of human systems within which an increasing majority of people throughout the world spend their working lives. This view melds the approach of systems analysis, as it has grown from its origin during World War II, with some of the theory and concepts of Policy Sciences which deal with policy analysis, extrarational variables, crosscultural variables, values analysis, political variables, and future studies.

The book, therefore, encompasses two rapidly developing, but interrelated scientific fields: policy sciences and systems sciences. Such melding allows for a theoretical framework and new practical concepts and techniques that bridge the quantitative—qualitative gap and increase the capability of systems analysis to be widely and more directly applicable to real world complex problems where it must demonstrate its effectiveness.

My central theme is that existing public or private human systems can be improved through the use of systems analysis. The determinations of whether we want them improved, whether constraints allow improvement, for whom they are being improved, and in what direction and degree improvements should be made are also areas where systems analysis can play an increasing role. Furthermore, systems concepts alert us to the appreciation that some of our systems should be redesigned to be improved or that entire new systems may be needed to replace an existing one or to fulfill new functions and needs emerging from our changing environments.

I define systems as a complex set of interacting elements. My meaning of the term "human systems" goes beyond the life sciences definition to combine the physical and technological connotations of the word "systems" with the social and human components of our public or private organizations. The English word "organizations" approximates my meaning; however, the word "systems" adds concepts of interaction, wholeness, interdependence, structure, process, feedback, inputs, and outputs that the word "organization" does not necessarily imply. Human systems, therefore, for the purpose of this book, mean any set of goal-oriented interdepen-

dent units incorporating people, organizational structure and some form of technology for control, administration, or production.

In today's complex world the gap between what people desire their systems to be
and the existing reality is growing and forecast to increase. Explanations for this
phenomena stem from the dynamics of change and the diversity of ratios of
change throughout the world in economic, social, cultural, scientific, and political
conditions.

The debate over the utility of systems analysis is not over and should continue as
a prod for critical self-analysis of those involved and a deterrent to hubris of systems analysts. However, to date Systems Analysis has provided the best available
set of analytical tools for describing systems, for identifying more efficient and
effective alternatives, and for implementing policy. Many areas where systems
analysis faltered on the limitations of its own overly rational methodologies in the
past can be effectively analyzed through the systems analysis—policy sciences mixture of concepts presented here. Pitfalls remain. Problems in complex human
systems cannot be manipulated in a Procrustean fashion to match available analytical tools. Usable solutions emerge when intelligent application of scientific knowledge, analytical models, and the judgement of experience are mixed. The inclusion
of extrarational components in our systems analysis does not make our efforts less
scientific than the alternative of analysis narrowly restricted to quantifiable models.
Quite the opposite is true. The scientific method also encourages a challenge to the
assumption that policy recommendations are not sensitive to arbitrarily excluded
extrarational variables.

Methodological tools provided here for Systems Analysis run the gamut from
behavioral research into what exists, through values research into what is preferred,
to normative research into what should be. Also included are research efforts to
demonstrate the three fundamental feasibilities—economic, technological, and
political—as difficult and ephemeral as they may be individually and considered
together. The pitfall of "For whom the analysis?" is a real world problem the
analyst must face. *Systems Analysis and Policy Sciences: Theory and Practice* does
not provide cookbook type answers for managers but does offer an analytical
framework to foster understanding of complexity rather than oversimplifying it.

In logic the book uses a combination of inductive and deductive reasoning: inductive in that problems for resolution initially emerge from specific systems functioning in their unique environment; deductive in that orientation for analysis
comes from the theory and models currently composing the paradigm of systems
analysis. The effective systems analyst must creatively apply the theory and models
of systems analysis to unique systems problems. The major dependent variable of
concern is the quality of the system as measured against standards consciously
selected.

In style and format the book is divided into four major parts: *I. Theory*—in the
belief that thought should precede analysis and action; *II. Practice*—illustrative
examples of the application of systems analysis theory to real world complex problems; *III. Synthesis*—a review and elaboration on the theory-practice mix; and
IV. Chapter Notes, References, and Bibliographic Essay—which will lead the reader

to relevant additional literature. Use of the bibliographic essay also reduces the need for lengthy footnotes or direct quotes, both of which I consider stumbling blocks to communication. Some readers of this book have recommended beginning with Part III. These four parts of the book are, themselves, systems components. Taken individually as subsystems they may have some value but the intent is for those parts to be synergistic and mutually interdependent. Each of the ten chapters of theory (Chapters 1-8, 19, and 20) provides an essential component of the theory of systems analysis presented. The totality of that theory will not be completely apparent until at least the initial eight chapters have been assimilated. To be entirely comprehensible the theory may need the support of the illustrative case studies provided in Part II or in forthcoming companion volumes or from the readers experience. Part III, Synthesis, draws together some of the strings that the reader may feel are loosely woven. Finally, for those who find some of the concepts foreign and inadequately explained, I have included a Glossary of Key Terms and Concepts and the Bibliographic Essays of Part IV where the reader will find a rich supplementary reading source.

For those who may find the absence of multiple detailed procedures frustrating I must explicate my strategy as being that of presenting a conceptual "How to . . ." book rather than a procedural "How to . . . " one. The book is intended primarily for the graduate level. My distinction between undergraduate and graduate education is that in undergraduate education the student is exposed to a segment of knowledge within one academic discipline. The educational process essentially "tells it like it is" and expects the student to absorb the material presented. In graduate education the student is expected to challenge—as well as utilize—existing knowledge within and across disciplines through the creative application of conceptual, analytical, and technological tools and techniques. Graduate education in Systems Analysis should involve explicit methodology, concentrated intellectual effort and creativity within the scientific method.

For the reader interested in conference room or classroom application I have designed all tables and figures to be suitable for directly converting into viewgraph transparencies for overhead projection. I have found this form of visual aid to be the most flexible educational vehicle—particularly where members of the audience hold English as a second language.

Systems Analysis and Policy Sciences has developed from personal experience, the experience of others, and my teaching of the "Systems Analysis" course in the University of Southern California's Master of Science in Systems Management Degree Program, which is provided at sixty study centers throughout the United States, The Pacific and Germany. The methodologies offered herein have, by 1980, had five years of real-world testing and have been instrumental in improvement through a wide spectrum of human systems. It is the first of a five-work series in systems analysis. The related forthcoming books: *Systems Evaluation; Values Analysis; Crisis Management;* and *Political Feasibility in Systems Analysis* will amplify, expand, and illustrate the theory and applications of this book.

I have attempted to avoid excessive systems analysis jargon as the book is meant for an international audience and transfer of complex English language into other languages is often difficult. In retrospect, I see I have not always been able to achieve that objective.

This book results from the direct and positive intellectual stimulus of two sources. First from my teacher, Dr. Yehezkel Dror of Hebrew University, Jerusalem, whose Policy Sciences theoretical and practical contributions to the improvement of the human condition have been immense. Secondly I wish to acknowledge the contribution of my USC students in the Pacific Area Study Centers during 1977, 1978, and 1979, some of whose systems analyses are summarized in Part II. In addition, I wish to thank: Dr. George Jones, Associate Professor of Systems Management, University of Southern California whose insightful critique of the entire manuscript resulted in many improvements—particularly to Chapter 8; Dr. John V. Grimaldi, Director of Degree Programs of USC's Institute of Safety and Systems Management who also critiqued the entire manuscript; my mother, Mrs. V. E. Spencer, provided a valuable review and correction from the perspective of her knowledge of English; and Sue, my wife, provided constant support throughout the three-year development of the book.

Systems Analysis and Policy Sciences is part of the increasing human systems improvement and design literature. The attempt has been to synthesize those elements of systems theory and those concepts applicable to the solution of problems through systems analysis conceived in a policy sciences framework. Systems analysis contributes to the qualitative improvement of human systems. For continual validation of that conviction and improvement of systems analysis as a discipline reader feedback is essential. I welcome your critical evaluation of the usefulness of this book in your life and work.

ROBERT M. KRONE, Ph.D.

Institute of Safety and Systems Management
University of Southern California
Los Angeles, California 90007
July 1980

Contents

Tables

All tables in the book provide extremely condensed statements, concepts and literature summaries. They are intended to be both explanatory and heuristic and to replace lengthy narrative otherwise required. I recognize the oversimplification and inadequate supporting information inherent in the technique but consider its advantages in providing a gestalt of the subject outweigh its disadvantages. The interested reader can pursue any viewgraph concept in detail through the references and the Bibliographic Essay, Part IV.

10 Traffic Management Systems

11 Nuclear Power Systems

14 International Defense Systems

15 Automation

16 Management Information Systems

Figures

17 Hydroacoustic Systems

19 The Role of the Systems Analyst

Glossary of Key Words and Concepts

As this book merges traditional systems analysis concepts of the 1940s to 1970s with newer concepts flowing from the policy sciences of the 1970s and 1980s, meaning of words and concepts can become confused and may be stumbling blocks to communication. This glossary is provided in an attempt to reduce that confusion, at least concerning my meaning of the key words and concepts used in the text.

Alternatives to the Systems Approach precedent, religious faith, cultural imperatives, social or political ideology, parochial suboptimization, power aggregation motives, historical lessons, nondirected counterbalancing pressures, mental metaphors and models, operational code assumptions, episodic proof, popular movement, evolution of human needs. Chapter 3.

Appreciation Set to become aware of a possible new set of relationships in the system. A Sir Geoffery Vickers concept.

Atomistic View of Science Scientific research into minute detail within a field. The opposite of interdisciplinary research.

Behavioral Research seeks to answer the questions of: What? When? How Much? How Many? Makes descriptive statements about things, events, relationships, and interactions. Observes, counts, and measures.

Control Knowledge knowledge about the use, development, and control of environmental knowledge and human knowledge.

Delphi Process the surveying of expert opinion on an issue or subject . . . can involve a complicated computer-based iterative process.

Dependent Variable That end result which achieves its formulation and quality from being acted upon by the independent variables. The effect of the cause-effect relationship in science. The variable for which an equation that contains more than one variable is solved.

Deterministic Analytical Models quantitative tools that are applicable to problems where there is only one state of the world assumed and where variables, constraints, and alternatives are—after making acceptable assumptions—known, definable, finite, and predictable with statistical confidence. Table 8-3 and Chapter 8.

Economic Feasibility one of the three fundamental feasibilities in systems analysis. It is the probability that economic resources will be available for the selected policy alternative.

Environmental Knowledge knowledge for the understanding, control and direction of the environment. This knowledge falls predominantly into the domain of the physical and natural sciences.

Evaluation the two-step process of determining quality of a system and judging that quality in comparison with standards. Chapter 7.

Explicit Knowledge knowledge gained through learning that is objective, logical, symbolic, or factual and can be articulated.

Extrarational human characteristics outside logical processes of reasoning such as experience, judgment, will, love, creativity, loyalty, or family commitment. Table 1-2.

Future Studies a component of policy sciences concentrating on normative research into alternative future outcomes and policies.

Human Knowledge knowledge for the understanding, control, and direction of individuals, groups, and society. This knowledge falls predominantly into the domain of the social, behavioral, and life sciences.

Human System a set of goal oriented interdependent units incorporating people, organizational structure, and some form of technology for control, administration, or production.

Independent Variable those system components that can be changed, or manipulated, to achieve desired results (i.e., impact on the dependent variable). The "cause" of the cause-effect relationship in science. A mathematical variable not dependent upon other variables.

Measurement the process of selecting and applying scales for use in judging the quality of a system. Chapter 7.

Meta-Policy policy concerning the process and structure of the policymaking system; or, policy about policymaking.

Models simplifications, simulations, and/or abstractions of reality. Models are used as tools for thinking or analysis and are the major devices humans use to represent, implement, or test the results of their theoretical thinking.

Normative Research seeks to confirm the hypothesis of what "should be" for the future by identifying and validating actions and means to achieve those prescribed ends.

Operations Research a term used synonymously with Systems Analysis in the 1940s to 1960s. Operations Research began as the application of mathematics and economics to World War II and postwar requirements for new defense systems. It has continued as a distinct subdiscipline of management science.

Paradigm a universally recognized and accepted body of knowledge within the scientific community. See Thomas Kuhn references.

Pareto Optimum the policy strategy conceived to produce results beneficial to all parties concerned and harmful to none. Abstracted from the writings of the 19th Century scholar. Vilfredo Pareto.

Policy Analysis a component of policy sciences involving research into public or private policymaking. The goal of policy analysis is the application of methods, models, and problem solving techniques to the identification or invention of policy alternatives that will achieve desired standards of quality.

Policy Sciences a complex set of policy-oriented disciplines and efforts. The main goal of policy sciences is the improvement of policymaking. The philosophical and theoretical concepts of policy sciences in this book are in consonance with the writings of Yehezkel Dror. Chapter 5.

Policy Strategies the guidelines, scope, postures, assumptions, and main directions to be followed by specific policies.

Political Feasibility one of the three fundamental feasibilities in systems analysis and a major variable of political oriented analysis. It is the probability that a policy proposal will be accepted for implementation.

Political Variables those having to do with the political process of who gets what, when, and how, such as consensus building, coalition maintenance, power aggregation, power shifts, and resource allocation. These variables exist in all human systems.

Primary Criterion the dependent variable, stated in overall terms, for determining quality of a system in evaluation. The total net desired output. See secondary criteria and Chapter 7.

Probabilistic Analytical Models those quantitative models where there are more than one states of the world and where each possible state must be estimated or defined to allow computation of the conditional outcomes of each decision alternative in each state. Alternatives may be numerous. Mathematics, statistical inference, and probability theory are used to reduce uncertainty within acceptable assumptions. See Table 8-4 and Chapters 8 and 9.

Problem a problem exists anytime the actual state and the desired state are not equal. Table 6-1.

Procrustean Metaphor the reduction and transformation of a complex reality by forcing it to fit a preconceived model. The metaphor stems from Greek mythology where the giant Procrustes hated nonuniformity to the point of cutting off wayfarer's feet so that they would fit his "Procrustean Bed."

Qualitative Tools in Systems Analysis those concepts, methods, and techniques designed to include the subjective and nonquantifiable variables in systems analysis. Table 8-1.

Quantitative Tools in Systems Analysis those using mathematics, statistics, computer sciences, and economics to analyze the objective variables. Chapter 8.

Science the segment of human effort dedicated to the increase and use of validatable physical and social knowledge.

Scientific Method in Systems Analysis the combination of assumptions and methodologies that, when followed, make systems analysis part of science as well as part of the art of management. See Chapter 1 and Table 3.1.

Secondary Criteria subcomponents of a system chosen for evaluation because they are considered, for good reasons, to be positively correlated with and more measurable than the primary criterion. See Primary Criterion and Chapter 7.

Standards of Evaluation preferred system qualities, selected for use in Step 2 of the evaluation process, to compare with the quality of the system determined in Step 1 of the evaluation process. Judgments of quality can only be made in comparison with standards.

System a complex set of interacting elements.

Systems Approach a scientific paradigm including, inter alia, concepts of wholeness, interdependence, organized complexity, and learning through feedback and crossfeed. Chapter 3.

Systems Analysis a set of techniques (qualitative, quantitative, and mixed) deriving its methodologies from the scientific method, systems philosophy and branches of various scientific disciplines dealing with the phenomenon of choice. Systems Analysis focuses on applied knowledge for the improvement or redesign of human systems or in the designing of entirely new systems that will more effectively reach goals. It is a mixture of art and science.

Systems Analysis Briefing a short presentation by the systems analyst to the decisionmaker summarizing the analysis. The goal of a systems analysis briefing is to instill an appreciation set, to educate and/or to be an input to the policymaking process.

Tacit Knowledge knowledge gained from living. Portions of tacit knowledge cannot be explicated. The source is Michael Polyani's "Theory of Personal Knowledge".

Technological Feasibility one of the three fundamental feasibilities in systems analysis. It is the probability that science and/or research and development will make available the necessary technology for the selected policy alternative.

Theory a set of interrelated scientific propositions of the "If . . . then . . ." and the "There is . . . " types. The set of propositions composing the theory is multi-dimensional and more stable than any one of the single dimension propositions. A theory explains reality and provides a framework for research, analysis, and prediction (e.g., a theory of systems analysis).

Values things or principles preferred. Values are standards of desirability and evaluation independent of specific situations.

Values Research or Values Analysis the investigation of values sources, values indicators, and values implications of individuals, groups or organizations in systems analysis.

Systems Analysis and
Policy Sciences

Part I

THEORY

Part I provides the theoretical and conceptual fundamentals and concepts for a systems analysis—policy sciences linkage. The lengthy historical controversy over the usefulness of systems analysis is mentioned but not retraced. Some social scientists may feel that the theory provided is too much asserted and too little defended in light of those old debates. To this I plead guilty. My main interest has been to satisfy a primary criterion of: "Does it have a high probability of usefulness toward the validation, improvement, or redesign of human systems?". I attempt to avoid self-deception or the deception of the reader by the requirement to extend that criterion to include the questions: "For whom does it work, when, within what definition of the system and in spite of what pitfalls?".

The reader may identify my personal values set emerging throughout the book. It is a combination of optimism and idealism hopefully tempered with an additional mix of realism. I will also state my personal values on the politics versus scientism issue since I have included political variables for analysis. Systems analysis and policy sciences are powerful instrumentalities for leadership—civil or military, public or private. They should be used to increase the effectiveness of the democratic political process not for avoidance or replacement of those processes. Good systems analysis may change political feasibility, as discussed in Chapter 5, but there is no politically pejorative hidden message intended to indicate any preferences for scientists to replace politicians as the final decisionmakers in society.

Chapters 1 and 2 set the scientific foundations for the concepts of the systems approach and systems analysis which are presented in Chapters 3 and 4. Policy sciences are described in Chapter 5 with some of the useful qualitative tools for the systems analyst. Chapters 6, 7, and 8 are the research methodology "how to . . ." prescriptions.

1

Knowledge and Science

Scientific knowledge as developed, structured, and applicable to systems analysis and policy sciences is discussed. Nine epistemological components of the theory and practice of systems analysis are provided. The scientific method of systems analysis is diagrammed with an explanation of why systems analysis is an art as well as a science.

Systems analysis focuses on applied knowledge for the improvement or redesign of our human systems or in the design of entirely new systems that will more effectively reach goals. The knowledge accumulated to date has emerged from the disciplines shown in Table 1.1. Knowledge for the making of intelligent and wise choices in complex systems does not fit neatly into any one of them. All knowledge possessed, the knowledge we need, and the knowledge that may become available to us can be conceptualized a different way into the following three categories:

1. *Environmental Knowledge.* Knowledge about the understanding, control, and direction of the environment. This knowledge falls predominantly into the domain of the physical and natural sciences.
2. *Human Knowledge.* Knowledge about the understanding, control, and direction of individuals, groups, and society. This knowledge falls predominantly into the domain of the social, behavioral, and life sciences.
3. *Control Knowledge.* Knowledge about the use of, and further development of, knowledge within the first two categories.[1]

Environmental knowledge has made the most tangible progress over centuries of scientific and technological research. Human scientific knowledge has increased much more slowly but made significant gains over the past 100 years. Control knowledge, or policymaking knowledge for effective use of knowledge about the environment and about humans acting within that environment, is less than 50 years old, remains poorly aggregated, underdeveloped, and is still marginally accepted as a separate category of scientific knowledge.

The "new interdisciplinary fields" listed in Table 1.1 all have underlying motivations and assumptions that place them, at least partially, within the control knowledge category. The identification of the third category as "control knowledge" is an acceptance of the fact that man has reached the end of "free fall" as Sir Geoffrey Vickers has described it, and must influence human destiny or have humanity

Table 1.1 *Knowledge Disciplines**

Traditional disciplines:	agriculture, anthropology, the arts, astronomy, ethics, engineering, history, law, life sciences, linguistics, mathematics, music, natural sciences, philosophy, theology
Social and behavioral science disciplines (post-18th century)	economics, education, international relations, political science, psychology, public administration, semantics, social anthropology, sociology
New interdisciplinary fields (post-19th century)	artificial intelligence, bionics, conflict studies, decision sciences, future studies, genetic control and engineering, geriatric studies, general systems theory, management sciences, world and national development studies, policy sciences, space sciences, strategic studies

*Due to the interdisiplinary nature of knowledge and the evolution of all disciplines any taxonomy is vulnerable to debate. The usefulness of this one lies in its demonstration of the relationship between discipline evolution and the phenomenon of choice. New knowledge is impacting on all disciplines and moving them toward broader scope, on the one hand, and also toward more detailed investigation of variables within the discipline, on the other hand.

swallowed up by the results of inability to do so. The three categories are interdependent and inseparable in the real world of complex human systems. That interdependence phenomenon is a 20th century realization about knowledge. Future knowledge increases must always face the test of implications on the environment and on humans. It can do this best through the evaluation methodologies developed and to be developed within the category of control knowledge.

The epistemological foundation, on which the theory and practice of systems analysis presented here is based, results from a mixture of the following components:

1. Rationalism tempered with the theory of personal knowledge of Michael Polyani[2] (see Table 1.2 for a condensation of that theory).

2. An idealism that assumes human system improvement is definable and achievable.[3]

3. A realism that sees a widening gap between needed real world systems improvement or design and organizational ability to apply available knowledge to effect that improvement.

4. Belief that values (see Chapter 5) and extrarational components such as judgment, will to achieve purpose, charismatic leadership, tacit knowledge, creativity, love, loyalty, freiendship, family ties, and serendipity are inextricably

Table 1.2 *Tacit and Explicit Knowledge**

Tacit Knowledge (Through Living)	Explicit Knowledge (Through Learning)
Unformulated	Articulated
Personal	Public
Experiential	Objective
Nonexplicated	Words, maps, formula, symbols
Gestalt processes involved	Logical
Extrarational component of culture	Articulate framework of culture
Guides commitment, explains expert knowledge of a science or art	Disciplines of learning
Basic to human understanding as is predominant means of transmittal of knowledge from one generation to the next	Facts
An indispensable part of all knowledge	Source of fallacies if divorced
Solves the paradox of Plato's Meno†	from tacit knowledge

"We know more than we can tell"

"What we know and can tell is accepted as true"

"We must believe to understand"

"Tacit knowing accounts for: (1) A valid knowledge of a problem; (2) The Scientist's capacity to pursue it, guided by his sense of approaching its solution, and (3) A valid anticipation of the yet indeterminate implications of the discovery arrived at in the end". 1966.

*Michael Polyani: *Science, Faith and Society* (1946); *Personal Knowledge* (1958); *The Study of Man* (1959); and *The Tacit Dimension* (1966).
†Plato stated that if all knowledge is explicit, then to search for the solution to a problem is an absurdity; for either you know what you are looking for, and then there is no problem; or you do not know and you cannot expect to find anything. His solution was that all discovery is a remembering of past lives.

linked with, and often more significant than, measurable variables in our systems.

5. A view that variations of both the deductive method of reasoning of Aristotle and the Scholastics and the inductive method of reasoning of Sir Francis Bacon and his followers are essential to systems analysis.[4]

6. A confidence that scientific knowledge can be increased by observation, description, classification, comparison, measurement, data aggregation, invention, experimentation, model application, theory application, and through extrarational processes. The scientific method by which human systems knowledge is increased, generated, applied, and analyzed is diagrammed in Table 1.3.

Table 1.3 *The Scientific Method in Systems Analysis*

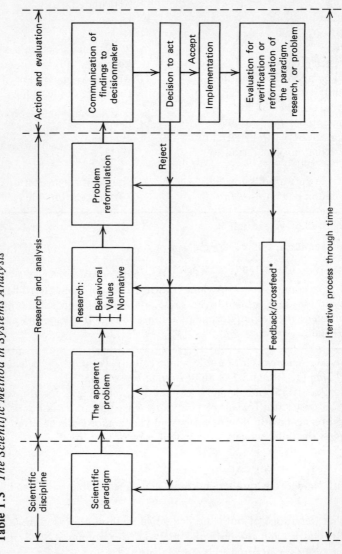

*Feedback is learning from the system. Crossfeed involves comparison with and learning from other systems.

7. A conviction that better crossfeed concerning other cultures' approaches to systems management can provide important insights and appreciations for the improvement of national and international public or private organizations.
8. Acceptance of the theory of Thomas S. Kuhn as the best explanation to date for how scientific knowledge evolves and progresses.[5]
9. The assumption that the central component of analysis can be either qualitative, quantitative, or mixed—which is the usual use.

According to Thomas Kuhn scientific knowledge grows within normal science through the aggregation of information consistent with a theory or through the replacement of one theory with a successor theory. He describes the process whereby normal science develops a paradigm, or universally recognized and accepted body of knowledge. This paradigm for a time provides solutions to problems for a community of scholars and practitioners. Over time, research and practice within this paradigm begins to reveal anomalous facts unaccounted for in the theory. This stimulates new scientific discoveries creating a crisis of credibility in the paradigm. A new theory emerges that precipitates a scientific revolution, or a nonaccumulative developmental episode in which an older paradigm is replaced in whole or in part by an incomplete new one—often with traumatic and long-range impacts on science, or scientists, and on society. There is ample historical evidence in the physical sciences to validate this theory and reflection upon the social sciences and the arts also reveals interesting insights when viewed through the Kuhn theory.

The scientific method, therefore, as applicable to systems analysis and shown in Table 1.3 has a conceptual beginning point in the prevailing scientific paradigm that can be transformed by the analytical process itself. As with any two-dimensional representation of something as complicated as the scientific method Table 1.3 is oversimplified and distorts reality. The scientific method is not necessarily linear, nor does it always flow neatly through the steps shown; however, the table is useful in providing a systems view of the essential components of the method. All are necessary and sufficient to bring a scientific orientation to the problems of systems improvement and design. The key one, from the standpoint of justifying our efforts as scientific, is that of evaluation for verification. That verification must occur for findings and recommendations of the specific problem-oriented systems analysis to be corroborated. Also essential is aggregation of knowledge about the ability of systems analysis theory and methods to achieve real world improvements over time. To avoid closed system analysis within unchallenged scientific paradigms the scientific community, as well as organizational and societal direction systems, must be involved in this latter macro validation process.

You will note, also, that I have identified three major categories of phenomena as relevant to the scientific method shown in Table 1.3. Those are "the scientific paradigm," "the research and analysis" category, and the "action and evaluation" category, which includes the processes of feedback and crossfeed. A major distinction between the scientific method in systems analysis and that of pure scientific research is the inclusion of a decisionmaker. Pure scientific research needs no decisionmaker other than the scientist's decision to conduct the research. Systems Analysis, however, passes through a decision point at some time. The decision, itself may or may not be based on scientific considerations—alone or in part. The analysis must have some scientific considerations—alone or in part. The analysis must have some scientific basis or it is outside the realms of systems analysis as presented here. That scientist-decisionmaker dilemma is resolved in an infinite number of ways in human systems. Due to space constraints that fascinating subject cannot be pursued. However, the "rejects" line running to the left of the "decision to act" box portrays the spectrum of possible impacts on the "problem reformulation," on the "research" on "the apparent problem" or even, in unusual cases, on the scientific paradigm itself, although paradigms are much more resistant to change than problems.

Those nine epistemological components, cited above, form the foundation of the theory and practice of systems analysis, which is developed in the chapters following. They are stated at the outset so the reader may use that foundation as a standard for evaluating and verifying what follows in terms of consistency and applicability.

I also feel it necessary to state my contention that Systems Analysis with explicit knowledge, and even the combination of explicit and tacit knowledge, is not always better than the pure judgment of leadership, cultural imperatives (taboo and folk lore), religious imperatives (faith or word of the prophet), or decisions flowing directly from political and social ideologies. The quality of the systems analysis is an important variable. Furthermore there are areas of human decision-making where systems analysis has been inappropriate (see Chapter 4). There will always be a value judgment required of the systems analyst and the decisionmaker concerning what mix of explicit scientific knowledge and extrarational variables is appropriate for the specific problem. My claim is, however, that systems analysis must be an available capability for decisionmakers in private or public human systems, in any culture, for the quality of policymaking to consistently meet whatever standards are selected.

Systems analysis conceived in a policy sciences framework is a combination of applied science and art. Science develops theory, models, and techniques, capable of validation, for handling unique problems of management and planning throughout a wide spectrum of government, business, military, industrial, agricultural, service, transportation, and communications systems. The art of systems analysis comes in the creative selection and blending of qualitative and quantitative tools needed for the problem. The science-art mixture applies the tools and evaluates the need for reformulation of the theories, models, and techniques when real world experience indicates that existing ones are inadequate.

2

Theory and Models

The essential role of models and theory in human progress through the creation, processing, and application of knowledge is briefly traced. Nine specific ways useful to systems analysts are listed. The theory of this book is placed in perspective to these nine contributions of theory within the scientific framework of the discussion of Chapter 1.

The ability to construct theory and models using symbolic language and thought provides the major distinction between humans and animals. Animals learn only from their own experience and through inherited instincts. Humans can learn from the experience of others as well as their own and even transcend experience through intellectual processes of storing, manipulating, and recalling symbols and abstractions of reality.

Scientific theory systematizes current and past knowledge and points the way toward new knowledge. Theory is a set of interrelated scientific propositions of the "if . . . then . . ." and the "there is . . ." types. The set of propositions composing the theory is multidimensional and more stable than any one of the single dimension propositions. A theory explains reality and provides a framework for research and prediction. The test of theory is that it must be capable of being shared, reproduced, operationalized, and verified repeatedly. Theorists need a capacity to test reality, to discriminate, to see causal relationships and patterns in chaotic data emerging from reality, to think abstractly, to conceptualize using both deductive and inductive logic, and to make understandable to others the results of their analytical thinking.

Models are the major devices humans use to represent, implement, or test the results of their theoretical thinking. Models are simplifications, simulations, and/or abstractions of reality. Table 2.1 summarizes the types, categories, functions, criteria, and pitfalls associated with the use of models in Systems Analysis.

There are at least nine different ways that the use of theory and models can be useful to systems analysts with regard to the generation, processing, and application of knowledge. Those nine ways are summarized in Table 2.2.[6]

The propositions provided in Chapters 3 through 8 are intended to expand the current theoretical foundation of Systems Analysis consistent with the requirements of theory presented above. In doing so I am aware of the further characteristic of new theory, that it constitutes a threat to those comfortable with, and committed to, the old paradigms. In the case of systems analysis theory, however, the anomalies of the "Operations Research (OR)" paradigm for human systems

Table 2.1 *Models*

A model is a tool for thinking and analysis.   A simplification, simulation and/or abstraction of reality; model conceptualization and construction is one of the most valuable human intellectual capabilities; model building may come before or after realty (space vehicle mockup or population growth models)

Models may be: Physical (model airplane); conceptual (world as a fixed pie); mathematical (expected value); pictorial (photograph); symbolic (The Cross); representational (map or clock); economic (Keynsian)

And categorized as: Behavioral (descriptive); normative (prescriptive); deterministic (one state of the world); probabilistic (more than one state); macro (world economic order); micro (atomic structure)

Model functions: Hypothesis testing; data organization and selection; system evaluation, analysis, simulation, prediction or improvement; identification or measurement of dimensions, variables, and relationships; a catalyst for creative observations, experiments, or alternatives; a learning device; risk or cost reductions

Model primary criterion: Must accurately simulate reality to be useful

Pitfalls of model use: Confusion of the model with reality (reification); excessive data collection; distortion of reality to fit the model (the Procrustean metaphor); tendancy to be overly rational; tendency to be closed where reality is dynamic and open; creation of opportunity costs through use of wrong model

---

analysis have already produced a high enough threshold of discomfort over a long enough period of time that alternative models for a systems analysis—policy sciences merger are now providing new tools within management science.

A theory should have implications greater than the sum of the uses of its individual components. The theory portions of this book are found in Chapters 1 through 8, 19, and 20. Those individual chapters contain components and models that form the set of interrelated scientific propositions which is multi-dimensional and more stable than any of the single components. For instance, the evaluation model of Chapter 7 is designed to be useful in itself as an applied science tool but is also an essential adjunct to the research methodologies of Chapters 6 and 8 which are themselves imbedded in the Policy Sciences and systems concepts of Chapters 3, 4, and 5.

As with a discrete system, an applied systems theory should be capable of partition into its parts then synthesizing those parts that focus on the system and its problems. The theory should help to bring applications, improvements, and designs into being with a higher probability of success than applications made in the absence of the theory. Theory provides guidelines, direction, and force; it can even have self-fulfilling prophecy characteristics.

The more real world documentation we have with applied theory the greater are its predictive strengths. This is why self-evaluation is so important for the systems management discipline. Our goal is not only to reduce human systems

Table 2.2 *The Role of Theory in the Creation, Processing, and Application of Knowledge*

Helps us to remember, store, and retrieve what we know through coding schemes (library systems, management information systems)

Can provide skills and guidelines for action (agricultural theory, theory of flight)

Provides empirical and pragmatic knowledge useful in goal setting, priority setting, and feasibility or pilot studies (marketing or city planning theory)

Organizes knowledge so we have more of its implications manifest (financial investment theory, trigonometric theory)

Produces new concepts, insights, and lables not previously recognized or formulated, through changes of scope, content, or form (general systems or policy sciences theory)

Provides knowledge of self-critical cognition to verify what we think we know (theories of management or national security)

Provides heuristic knowledge to suggest, stimulate, direct and design search routines for experiments and investigation (systems analysis theory or the theory of relativity)

Promotes awareness of what we want—values and goals—as well as what we know (social and political theories, theories of knowledge, aesthetics)

Can contribute to wisdom through the ability to identify: which values ensembles are best for the future; which theories and goals are worth the commitment of knowledge, time, and resources; which patterns of goals are conflicting or destructive; or which knowledge should be enhanced (values theory, theories of human motivation and improvement, theories of control knowledge utilization)

---

failures and increase their effectiveness but also to know what theories, what models, what management tools, and what techniques were employed and with what degree of success. In the absence of that knowledge the negative impacts of trial and error, discussed in Chapter 5, can be expected to increase with our increasing human system complexity.

A rich source of literature on theory and models relevant to Systems Analysis and systems management can be found in the previous volumes of the Wiley Series on Systems Engineering and Analysis, edited by Harold Chestnut. For example Chestnut, himself, has provided the basic modeling and simulation explanation (1965); Bernard H. Rudwick (1969) has described the use of models in quantitative systems analysis of military weapon systems; Hawn and Shapiro (1966) have presented statistical models in engineering; the Ralph F. Miles, Jr. collection of writings on systems concepts (1973) includes decision and quantitative models; Wilton P. Chase (1974) gives us the basic model for system project management; Gerald M. Weinberg (1975) provided an analysis of the theory and laws inherent in general systems theory to 1975; Dorny (1976) discusses models and optimization from the perspective of vector space; John N. Warfield (1976) has tackled theories and models in the policy process; John de S. Coutinho (1977) has investigated model types deficiencies, maintainability, and reliability; and House and McLeod (1978) looked at large-scale models for policy evaluation. Other theory and models sources can be traced through the literature of the General Systems Movement and in the writings on the systems approach, systems concepts, systems analysis, and systems management cited in the Bibliographic Essay of Part IV for this and succeeding chapters.

This book is part of the conceptual evolution of systems analysis and management science toward policy sciences and presents a normative theory that may have General Systems Theory application but is designed primarily as a management oriented applied set of tools for real-world problems.[7] In Part III, Synthesis, Chapters 19 and 20, I return to the subject of linking theoretical and practical systems knowledge for analyzing problems and achieving improvements in the complex world of our human systems.

3

Systems Concepts

The useful concepts and potential pitfalls of systems thinking are provided plus fourteen alternative approaches to the solution of problems that fall between the atomistic and systems views of the world. A somewhat fanciful possibility for the new scientific paradigm that might replace the systems approach in the 21st century is suggested.

One of the pioneer philosophers and writers about the systems approach, C. W. Churchman of the University of California at Berkeley, ends his provocative book *The Systems Approach* with the final statement: "The systems approach is not a bad idea."[8] That was 12 years ago and a better idea has not yet emerged. No doubt one will in the future and the systems approach will be replaced by a new scientific paradigm.

Conjecture about what that paradigm might be is a legitimate focus of future studies, one of the major components of Policy Sciences. I will here provide only one illustrative alternative future for the elements of that new paradigm, which I label "Reality 21." Under the Reality 21 paradigm our analytical capability would have progressed to the point where we would not have to theorize and conceptualize systems approaches to simulate reality. Our capability for empirical knowledge collection, reproduction, aggregation, and manipulation would offer us a real-time view of reality. Whatever segment of physical, social, cultural, economic, technological, health, biological, transportation, communications, meterological, energy, or agricultural life in the world policymakers or scientists were interested in analyzing would be available on command. The medium of reproduction might be three-dimensional laser-generated hologram images superimposed with desired data capable of sensitivity analysis in any dimension. With another command instant outcome arrays for a spectrum of normative probabilistically accurate world states would emerge for the future—each being the dependent variable resulting from allocations of designated independent variables of material and intellectual resources. Furthermore there would exist the capability for rapid Delphi surveying of every world citizen and organizational entity to obtain values analyses concerning which future states were preferred, with what intensity and in what manner those individuals and organizations were prepared to contribute to achieve the preferred future states.

Alas or, fortunately, depending upon one's feelings of how the political and social world might look under those conditions—Reality 21 is not a reality in 1980.

I will leave that concept for a later book and deal with this thing called the systems approach of the 20th Century.[9] It remains the best idea currently available.

Table 3.1 contains the useful concepts of the systems approach abstracted from the literature and my own experience. Table 3.2 summarizes the potential pitfalls of the systems approach arrived at in the same manner. Although recorded history is filled with individual examples of systems thinking and analysis within the traditional scientific disciplines, it was still possible for Robert S. Lynd to accurately state in 1939: "Modern science tends to be atomistic. Its drive is to isolate smaller and smaller variables and to study these in the greatest possible detail with the aid of minute controls."[10] World War II precipitated the beginnings of the systems approach to problems of warfare. Systems theory, philosophy, and semantics have grown since then to the point where in 1972 Ervin Laszlo described it as "the emergent and preferable scientific view of the world"[11] and one that substitutes concern for relationships within integrated and organized wholes for isolated concern with events and facts narrowly bound by knowledge disciplines. Although systems analysis is dependent upon system theory and philosophy for its genesis and much of its world view, the systems analyst must be careful not to completely succumb to either the fast talking stranger from systems philosophy (the generalist) or from atomistic science (the specialist).

Table 3.1   *The Systems Approach—Useful Concepts*

Definition: A complex set of interacting elements

Systems are characterized by: Wholeness; organized complexity; interdependence; reciprocal dependence (an action within a system causes other actions); dynamics (systems grow, alter, decay, die over time and through interventions); components of inputs, people, structure, process, outputs, and boundaries; interchanges with their environments; equifinality (from many beginning states a final state may be reached, and from one beginning many end states may be reached); Gestalt phenomenon (the whole is greater than the sum of its parts); health and/or pathologies.

Systems have functions of: System maintenance, policymaking, planning, goal setting, memory, control, learning, feedback, crossfeed, purpose, and unifying force

Systems can be evaluated through the use of criteria and standards

The systems approach: Provides understanding and comparisons within and between systems; encourages simultaneous research into different parts of the system; leads to awareness of the hierarchical nature of living and natural systems; creates knowledge; helps recognize what you do not know; asks different (and often better) questions than alternative approaches to problem solving; forces consideration of coordination, control, level of analysis and implementation as well as goals and problem solution; stimulates innovation; suggests a means-ends investigation; becomes a variable in the system product quality; is intellectually challenging and increasingly relevant; "Is not a bad idea"[8]

Table 3.2   *The Systems Approach—Pitfalls**

Neglecting to consider the following potential pitfalls of the systems approach can lead to gross conceptual errors that cause incorrect findings: Interchanging the abstracted system with reality; subsystem sensitivity to system assumptions ("What's good for the marketing department is good for the company"); problem of individual benefits versus system benefits (the "Who is the client?" problem where improving the system may mean firing people); requirement but inability to study adequately more than one part of the system due to complexity (a dilemma since the systems approach is the most useful one available for handling complexity); the part selected for analysis may not be the most important one (poor prioritizing due to quandary of "I have defined the system now where do I start?"); neglecting history; relying on historical metaphors; shortage of competent analysts; knowledge adequate for only a superficial analysis; bias toward rational thinking; excess analysis; using the systems approach for *Nonapplicable situations*

Nonapplicable situations include: When aim is to force a rapid change through shock; when goals are destructively negative; when leadership aim is power recruitment, coalition maintenance, or political consensus; when social-political ideology, faith, or family loyalty—not reason—are the prime movers; pure scientific research; the arts

---

*See Table 4–3 for Systems Analysis Pitfalls and Chapter 3 for alternative problem solving approaches to the systems approach.

Systems concepts are certainly necessary for systems analysis—but not sufficient. Bread and butter type empirical data is collected and tools are forged by experts committing their energies toward ever more rigorous and detailed knowledge of the behavior of the components of systems as well as the behavior of the systems as a whole. Difficult? Yes, but it helps to explain why the use of systems analysis teams and project management are on the increase.

There are alternative approaches to the solution of problems between the atomistic view of the world and the systems view of the world. Systems analysts must recognize that policy and action also result directly from the following:

1. Precedent of past policy or action.
2. Social and political ideology.
3. Power aggregation motives (e.g., political consensus).
4. Cultural imperatives such as tradition, superstition, and behavioral taboos.
5. Gradual evolution and integration of human needs into policymaking (as in the Third World rising expectations since World War II).
6. The suboptimizing attitude of "What's good for the production division is good for the company."
7. Historical lessons (e.g., "No more Vietnams").
8. Episodic proof based on one incident.

9. Nondirected counterbalancing pressures such as in the Middle East.

10. Mental metaphors and models (the world is a "fixed pie" of resources).

11. Operational code assumptions (relevant image of the opponent's intentions on which to base counter-strategies).

12. Popular movements (e.g., Ralph Nader and the U.S. consumer movement).

13. Religious faith, including messianic views of destiny.

14. Charismatic or authoritative leadership.

I do not maintain that systems analysis in a policy sciences framework is always preferable to the above modes used individually or in combinations. For example, an appropriate mental metaphor is preferable to poor systems analysis. I do contend, however, that policymaking quality will improve through wise application of systems analysis and, even further, that the absence of any systems analysis capability will allow deterioration in complex systems.

4

Systems Analysis

Systems analysis and systems are defined. The two distinct branches of development of systems analysis—the operations research branch and the academic disciplines branch—are briefly traced. The assumptions of utility and levels of analysis of systems analysis are provided. Systems analysis pitfalls to complement and continue the pitfalls of the systems approach provided in Table 3.2 are listed in Table 4.3. This chapter sets the stage for discussing the qualitative and quantitative methodological concepts and techniques of Chapters 5 through 8.

Systems analysis is a set of techniques—qualitative, quantitative, and mixed—deriving its methodologies from the scientific method, systems philosophy, and branches of various scientific disciplines dealing with the phenomenon of choice. The application of systems analysis aims at the improvement of public and private complex human systems. Systems analysis incorporates both explanatory and prescriptive methodologies.

A system is a complex set of interacting elements. The word "system" has provoked criticism, even hostility, in some areas primarily due to mistaken inferences of antihuman biases or due to the conflict which emerged from the oversell of systems analysis in the U.S. Government in the 1960s. Since then content analysis of professional journals, university texts, conference agendas, newspapers, telephone books, and advertising would reveal an increasing reliance on the word "system." Systems philosophers have yet to invent a better generic word to describe the natural and living phenomena under discussion. No doubt it is too much to expect them to do so within normal systems sciences—that must wait for the future paradigm which will replace the systems view of the world.

Until then it must be understood that systems are what people define them to be and what nature has bequeathed. Organizations are systems but not all systems are organizations. World energy consumption and production has been viewed as a system. My office can be considered a system and is definitely exhibiting increasing entropy. National governments are, inter alia, systems. Airplanes I used to fly were full of electronic, hydraulic, pneumatic, mechanical subsystems composing one integrated flying system, which was, itself, a subsystem of a human system called a squadron containing aircraft, pilots, maintenance, and administrative personnel. This example could be pursued to the United States Air Force level as a perfect example of the hierarchical nature of systems. Eric Berne's invention of Transactional Analysis (TA) has been called a "whole psychotherapeutic system."[12]

It does not follow that since almost anything can be defined as a system that the term is useless. The concepts identified in Table 3.1 lead to an analytical world view significantly different and in many ways superior to its predecessors.

During the three decades 1940 to 1970, systems analysis took two distinct branches of development. First was the application of mathematics and economics to the World War II and postwar requirements for new defense systems. Other names alternatively used in the 1940s and 1950s were "operations analysis," "operations research," "systems engineering," and "cost-effectiveness analysis." The term "systems analysis" and a great many of the influential analysts who attracted the attention of the Eisenhower and Kennedy Administrations in the 1950s and 1960s, came from the RAND (Research and Development) Corporation in Santa Monica, California.[13]

Both systems analysis and its conceptual follow-on, the planning programming budgeting systems (PPBS) officially moved from RAND to Washington under the auspices of President Kennedy's Secretary of Defense, Robert McNamara following the 1960 election. The three major categories of PPBS were: (1) long range budgeting; (2) a management information system to keep track of expenditures and programs; and (3) systems analysis.[14] On the basis of experience in the Department of Defense, President Lyndon Johnson directed that PPBS be implemented in all Federal Departments and Agencies of the U.S. Government on August 26, 1965. The story of why that directive was never fully implemented due to the controversy (including Congressional Hearings) over uses and abuses of systems analysis and PPBS does not make as exciting reading as the subsequent Pentagon Papers or the Watergate Tapes, but is relevant and recommended to the reader in the bureaucratic birth and adolescence of systems analysis in the United States.*[15]

During this same time period portions of the private business sector were utilizing systems analysis techniques for the improvement of efficiency and effectiveness in transportation systems, communications systems, computer systems, health care systems, and perhaps the best example the beginnings of the unmanned and manned space programs. The distinctly American phenomena of the think tank utilizing systems analysis and policy analysis proliferated from RAND to private and semiprivate research organizations across the country and to other countries.[16] I cite these few historical details merely to indicate the significance to public policymaking in the United States of this branch of systems analysis.

The second branch of development of systems analysis emerged from University-associated research and teaching in the form of systems theorizing in many academic disciplines beginning with biology and mathematics (particularly cybernetics) and extending to engineering, communications, General Systems Theory, economics, political systems, international systems, management systems, ecosystems,

---

*The controversy in the 1966–1968 period became so heated that one of Richard Nixon's first Executive Orders after assuming the Presidency in January 1969 directed deemphasis of systems analysis within the Department of Defense.

psychological and psychiatric systems, and education systems. These two branches of development—the economically rational think tank systems analysis and the theoretical and social orientation of academic systems analysis—followed largely separate paths until the 1970s when events and the efforts and writings of Yehezkel Dror and other policy scientists managed to demonstrate the usefulness of taking a systems approach to the further development of systems analysis and policy analysis for the improvement of policymaking systems.[17]

My guess is that there are as many definitions for systems analysis as there are systems analysts. The frustrating thing about them all—mine at the beginning of this chapter included—is that none answer the question of "What is it?" in other than generalities. A more satisfactory line of investigation is one that looks at assumptions, processes, methodologies, techniques, and applications—an approach I shall take for the remainder of the book.

All definitions of systems analysis assume that its utilization will result in:

1. More fully considered alternative choices identified for decisionmakers.
2. More efficient and effective utilization of scarce, expensive human and material resources.
3. Cheaper and/or better achievement of goals.
4. Improved policymaking—certainly for rational resource allocation problems and policy implementation problems, but also hopefully for goal setting and problems with social, cultural and political components.

That list of assumptions in itself implies a wide spectrum of analysis capabilities and levels of analysis. I have identified 10 levels of systems analysis divided into the two categories of "meta-analyses" (or analysis about the problem of analysis, Table 4.1) and "system analyses" (Table 4.2) reflecting a scope of capabilities from doing everything to doing nothing. Many of these levels are illustrated in the case studies of Part II.

Table 4.3, "Systems Analysis—Pitfalls" complements the systems concepts pitfalls provided in Table 3.2 by citing specific potential pitfalls for the system analyst. Neglecting to consider these potential pitfalls runs the risk of disastrous, or at a minimum, useless analysis. Most are self-explanatory and many will be singled out for discussion in the pages following.

Table 4.1 *Levels of Systems Analysis**

Meta-analyses:

Conceptualize theory for systems analysis

Do a study to identify resources, knowledge, and time needed for a system analysis

Design a systems analysis approach or model (quantitative, qualitative, or mixed) appropriate for analysis of a category of problems

---

*See Table 4-2 for systems analysis levels 4 through 10

Table 4.2 *Levels of Systems Analysis**

Systems analyses:

Do a complete systems analysis on a whole system, including behavioral research (what is); values research (what is preferred); and normative research (what should be), involving feasible alternatives and preparation of written reports and briefing medium for decisionmakers

Do a systems analysis, same as above, on a manageable subsystem because of interest in that subsystem or for a pilot study or experiment

Prescribe a system improvement, design, or redesign to resolve a specific problem, involving mostly normative research and assuming possession of in-house explicit and tacit knowledge in lieu of extensive behavioral and values research

Do behavioral research to discover what exists and to clarify specific problems in a confusing situation with poor data

Accomplish a values analysis to answer questions of: "Who prefers what?"; "Who are we?"; "How should preferences be prioritized?"

Create a new systems analysis model to evaluate or replace a model previously adopted with marginal or poor results and to identify improvements considering lessons learned from the earlier failure.

Do nothing under the assumptions that: the present quality of the system is good enough; or, future events will resolve the problem without formal analysis; or, the constraints on your analysis capability preclude systems improvement or redesign

---

*See Table 4-1 for systems analysis levels 1, 2, and 3 (meta-analyses)

Six types of problems have been outside the domain of effective applicability for systems analysis with its economically rational models and tools:

1. Problems with high political content where goals are opinion consensus, coalition maintenance, power recruitment, or nation–building.
2. Problems with high social implications. For example systems analysis has been extremely successful in improving and designing efficient logistics systems but not with issues of racial segregation and discrimination.
3. Problems where extrarational processes play important roles in the decision.
4. Problems where values (things preferred) must be weighted along side the value (utility, price) for alternative selection.
5. Problems where the strategy desired calls for radical or shock treatment of the system rather than a balanced consideration of all system components.
6. Problems of policy implementation when new institutions–not working within existing institutions–becomes necessary.

These areas fall either into the domain of policy sciences–which I discuss in the next chapter–or outside science entirely. The mistakes of some systems analysts

Table 4.3 *Systems Analysis Pitfalls*

A Partial Checklist: Overlooking importance of conceptualizing the effort prior to selection of techniques; overestimating and overselling analytical performance capability and/or underestimating time and effort available versus needed; adapting the problem to fit the available methodology (the Procrustian solution to complexity); failing to reality test through model fixation; assuming that no quantification is necessary; assuming that quantification is all that is necessary; confusing values (preferences) with value (utility); assuming the analysis process is values neutral thus failing to explicate values, either your own or the system's; failing to identify noncompromisable absolute values; using wrong techniques, models, criteria, or standards; neglecting applicable feedback or crossfeed inputs; assuming that analysis must always be costly; depending too much on technology; neglecting past and future time impacts; group-thinking analysis team or institution; overlooking unique system features; making good incremental decisions that lead to bad end results; Neglecting the fact that systems analysis is an art (extrarational expertise) as well as a science (rational expertise); forgetting risk analysis . . . limiting alternative search too much; ignoring one of the three feasibilities (economic, technological, or political); . . . forming a poor image of the adversary (how would he attack your analysis or counteract your actions recommended?); Forgetting that there are problems outside the domain of systems analysis dealing with faith, politics, communications, and culture; Failing to communicate analysis understandably to decisionmaker(s)

---

of the 1950s, 1960s, and 1970s was to assume, or fervently hope, that the real world which included the above six categories of problems could be analyzed and molded with economic cost/benefit tools alone.

The truly significant contribution of systems analysts in that time period was to permanently instill systems thinking into the appreciation set of policymakers and planners and to continue the development of new tools to handle the complexity of the real world. The quantitative tools developed by systems analysts will be covered after a look at the qualitative tools and methodologies of policy sciences in Chapters 5, 6, and 7.

5

Policy Sciences
and Qualitative Tools

The reader is given the short history of the development of policy sciences plus its dimensions and concepts. Some qualitative analytical tools are discussed including policy strategies; policymaking models; values analysis; consideration of extrarational processes; political variables; cross-cultural variables; and future studies. The requirement for the systems analyst to include these largely nonquantifiable social and psychological phenomena is acknowledged to provide a quantum jump in complexity but a correspondingly significant increase in relevance to the world of human systems. Many of the case studies in Part II utilize one or more qualitative tools of policy sciences.*

A single operational concept of policy sciences has not yet emerged and—because of the cross-disciplinary nature of knowledge involved in the formulation, selection, analysis, and implementation of policy—may not emerge with any large degree of consensus in the near future. Nevertheless, the fundamental dimensions of policy sciences are taking shape. I will describe the background and dimensions of policy sciences, drawing on the works of Yehezkel Dror of Hebrew University, Jerusalem as the most energetic and articulate advocate and educator of policy sciences.[18]

The movement of Harold Lasswell's 1951[19] idea for policy sciences to general acceptance in the late 1970s was largely due to Dror's personal commitment. During his 1968 to 1971 residence in the United States he published his trilogy of policy sciences works: *Public Policymaking Reexamined (1968),* which presented a general theory of policymaking as a basis for policy sciences; *Ventures in Policy Sciences (1971);* and *Design for Policy Science (1971).* In those same years Dror produced a large number of policy sciences monographs in his capacity as Senior Staff Member, the RAND Corporation (Santa Monica); encouraged various American universities to begin programs in policy sciences; taught courses at UCLA and USC in Los Angeles; began, with E. S. Quade at RAND, a new journal called *Policy Sciences;* encouraged Harold Lasswell to write his "twenty-years afterwards" view of the status of policy sciences[20]; organized and directed the first policy sciences

*Permission to use edited portions of my article "Policy Sciences and Civil Military Systems" *Journal of Political and Military Sociology,* Vol. 3, No. 1 (Spring 1975), pp. 71–84 for this discussion was kindly granted by the *JPMS.*

panel at an American Association for the Advancement of Science 1969 convention; encouraged other panels at American Political Science Association conventions, and laid the groundwork for a policy sciences book series by American Elsevier, under his editorship. In addition, he has participated since 1972 on the editorial board for the *Policy Studies Journal*[21] ; conducts periodic international workshops in policy analysis for senior decisionmakers, and is influential in improvement of Israeli policymaking systems. The following very condensed description of policy sciences, then, has its conceptual roots in the works and teachings of Yehezkel Dror.

THE EMERGENCE OF POLICY SCIENCES

There was little movement in the 1950s in response to Lasswell's proposal for Policy Sciences. A number of statistical, mathematics, and economic tools and techniques for decision analysis were independently developed and improved, government decision-making studies increased, and problems of choice were investigated—primarily in nuclear strategy, business, and individual choice. Other developments in common with systems analysis and policy sciences, such as systems theory and cybernetics were being developed in the biological and mathematical sciences. Management sciences, a parallel and overlapping discipline with systems analysis, was reaching a postwar take-off point.

The 1960s saw the acceptance of policy analysis on a wide scale throughout the U. S. governmental agencies. Even more important from the standpoint of policy sciences was the emergence of the uniquely American phenomenon of research centers organized to devote concentrated professional attention to problems of national defense, economic planning, technological development, space exploration, urban growth, education, poverty, and other social problems.[22] Systems analysis was actually a subcomponent of policy sciences although both developments were occurring concurrently.

The following converging causative factors produced a sudden upsurge of interest in policy sciences in the 1965 to 1970 period: Increasing public concern over specific policy issues such as war, poverty, crime, race relations, pollution, and transportation; concurrent increased awareness by the U.S. intellectual community and by citizens that solutions to problems were often of unsatisfactory quality; frustration over the comparison of achievements in the physical, biological, and space sciences with the apparent inability to solve social problems; student-generated crises on university campuses over the relevance of academic course content to obvious local, national, and international problems and over policies of the various establishments to resolve these problems; requirements for crisis policymaking (e.g., in reaction to Sputnik, urban riots, campus unrest, and Soviet missiles in Cuba); and an increasing number of scholars, scientists, and policy advisors with the unique combination of necessary intellectual prerequisites, practical experience and commitment toward the establishment of policy sciences as a distinct interdiscipline.

DIMENSIONS AND CONCEPTS OF POLICY SCIENCES

Policy Sciences is concerned with the application of knowledge and creativity to better policymaking. Its boundaries are imprecise due to the knowledge requirements that cut across the three categories of knowledge shown in Table 1.1. Policy sciences is a distinctive new interdiscipline not a replacement for existing ones. If any generalization can be made with regard to the dimensions of policy sciences, it would be that policy sciences encompasses any pure or applied research effort involving the phenomenon and context of choice and dealing with the policymaking system. During the 1960s policy sciences made its greatest impact on the public policymaking of national governments. In the 1970s the private sectors also became aware of its potential.

Philosophically, policy sciences views the evolution of mankind as having reached that point in history where man must shape his future or be swallowed up by the undesirable results of his own inability to do so.[23] Therefore, the improved quality of policymaking and the destiny of the human race are, for the first time, directly linked. A further assumption is that the new endeavor of Policy Sciences is essential because of the inability of contemporary sciences to meet urgent policymaking needs or to adequately address the problem of the interrelationships between scientific knowledge, political power, and social implications of power and knowledge.

The motivation of those laboring in the various policy sciences areas can be explained through the assumptions inherent in the following syllogism:

1. The ways in which policy decisions are made—especially trial and error—have changed relatively little through the ages. The decision to try a specific solution stems from a basic set of values within which some policy alternatives are considered and one selected. The error is then observed. Feedback may occur and result in system improvement.
2. Society's problems are increasing in complexity, interdependence, and intensity at an increasing rate.
3. Therefore, without increases in policymaking knowledge and the incorporation of this knowledge into the policymaking system it can be expected that the same or higher percent of future policies will be bad ones (i.e., not result in achieving minimum necessary goals within acceptable values) as in the past; however, in the future, impacts of bad policies will be worse and possibly irreversible.

A summary of that syllogism is: without improved policymaking the future holds even larger problems for society.

COMPONENTS OF POLICY SCIENCES

Policy sciences involve research toward the understanding and improvement of human systems through qualitative and quantitative methodologies; a focus on the policymaking systems level under the assumption that discrete policy improvement will flow from improved policymaking systems; the capacity to analyze values issues; acceptance of tacit knowledge (see Table 1.2) and extrarational processes (see Table 5.3) as important sources of knowledge; the encouragement of organized creativity for policymaking system redesign and policy alternative inventions; time sensitivity (e.g., recognition of historical legacies, potential future developments and the impact of time); and explicit consideration of political and cultural variables.

Five policy sciences areas for methodological focus are: (1) policy strategies; (2) policy analysis; (3) policymaking system improvement; (4) evaluation; and (5) policy sciences advancement. Space constraints preclude exploration of all potential benefits of linking the macro and social orientation of policy sciences to the relatively micro and economically rational orientation of systems analysis. This book is designed to be a heuristic aid to stimulate further investigation of those benefits.

Chapter 7 addresses evaluation exclusively. Requirements for the advancement of policy sciences and the implications of policy sciences for both science and for politics are best presented by Yehezkel Dror in his 1971 book, *Design for Policy Sciences,* Parts IV and V. For the remainder of this Chapter, I shall discuss only the methodological areas of policy strategies and policy analysis. Policymaking system improvement is perhaps the concept of greatest overlap between systems analysis and policy sciences and permeates both the theory and the practice of systems analysis presented here. The differing approaches to systems improvement of policy sciences and systems analysis are due to methodology orientation, levels of analysis, and somewhat to systems semantics. Whereas policymaking system improvement is both a major subject area of policy sciences and a normative goal of policy sciences not all of systems analysis deals with the policymaking system or its improvement. Systems improvement can involve changes in input, personnel, structure, process, output, or environment of the system. Much of systems analysis deals with the efficient, effective and orderly utilization of resources allocated by a policymaking system. Other portions of systems analysis do deal directly with policy formulation and the policymaking systems and it is those portions, particularly, that are given conceptual vitality by policy sciences.

POLICY STRATEGIES

Policy strategies include the guidelines, scope, postures, assumptions, and main directions to be followed by specific policies. Table 5.1 lists some major policy strategy options often neglected in policymaking and systems analysis. Decisionmakers hold tacit theories concerning policy strategies which, if not subjected to explicit scrutiny and reevaluation overlook complete ranges of potential alternatives. I will return to policy strategies through references in the case studies of Part II.

POLICY ANALYSIS

By the last half of the 1970s a large number of policy-oriented degree programs, courses, research centers, and professional journals had sprung up on American

Table 5.1 *Policy Strategies*

Policy scope: The identification of which system elements should be included in policy analysis or change and to what degree, to include the no-decision model which allows environmental variables to determine the outcome (international currency floating)

Goals strategy: To identify specific goals or the goal to build capacities for future policymakers to achieve as yet undefined goals (train to match job description or educate for future policymaking capability); positive or negative goal statements (conservation versus production to meet energy consumption goals); The Pareto Optimum (the strategic goal of searching for a policy which will result in net benefits to all parties)

Strategy options within spectrums of: Incremental improvement to radical redesign or system termination; short to long range target time preference (present or future discounting); low risk-certainty-security versus high risk-uncertainty-low security-high payoff; optimization to suboptimization to min-avoidance (the strategy of moving from worst alternative to worst-plus-one—President Sadat: "No more war!" to Israeli Knesset, 1977; centralization to decentralization; balanced policy to shocking the target system; cooperation to conflict toward adversaries

Mixes within the strategy spectrums and alternatives above

---

campuses due to the causative factors that gave policy sciences its stimulus plus the widespread student discontent in the 1960s and early 1970s over the non-relevance of curricula to obvious societal problems.

Policy analysis combines the quantitative and economic methodologies of systems analysis (see Chapter 8) with some of the qualitative methodologies and considerations of policy sciences. The goal of policy analysis is the application of methods, models, and problem-solving techniques to the identification or invention of policy alternatives which will achieve desired standards of quality (see Chapter 7).[24]

I have included six areas of policy analysis for discussion: (1) policymaking models; (2) values analysis; (3) extrarational variables consideration; (4) political feasibility; (5) cross-cultural variables; and (6) future studies. A priori exclusion of any of these areas from policy analysis or Systems Analysis can lead to poorly developed policy alternatives.

POLICYMAKING MODELS

All decisionmakers carry their own mental models of how policymaking should be accomplished although not all decisionmakers can explicate those models. I have identified in Table 5.2 seven pure types of policymaking models that are and have been used for public and private policymaking. There are explanatory and normative advantages to the systems analyst of such a typology. Having identified

Table 5.2 *Policymaking Models: A Typology*

The rational model. Assumes that human action approximates, or should approximate, rationality; or can be usefully explained as if it did; typical phases include identifying problem, gathering data, listing all possible solutions, testing all solutions, selecting the best, and acting; ignores extrarational variables; . . . has its roots in 19th and 20th century Western civilization *(study everything to x^2)* *infinite time & money*

The economically rational model. Uses the rational model in so far as it is economical to do so; considers extrarational variables as unavoidable evils and prefers to ignore them; has a strong tendency to quantify; was an early user of systems analysis (cost/benefit, operations research); . . . is the basis of much of business theory, economics, some political theory and defense theory; PPBS; ZBB *(benefit of payback = or > than expenditure)*

The incremental change model. Holds that policy is, and should be, made through slow evolution and cautious change; . . . that objectives should be adjusted to feasible means; that optimal quality is utopian; that satisfactory quality is the best obtainable; is conservative; is skeptical about human ability to radically change the future; is used by most systems *(in reality change will be only accepted in increments)*

The sequential decision model. Used where insufficient knowledge or consensus is available; commences with simultaneous and redundant approaches, or decides on phase I, assuming new knowledge or consensus will emerge for subsequent phases; then selects preferred alternative as knowledge and feasibility emerge; offers alternative to decision under extreme uncertainty or no decision *1. proceed along w/o final decision (conscience) 2. simultaneous approaches (decision tree)*

The extrarational model. Policy flows from extrarational characteristics of system without a rational link; can be irrational *(opting, to go with things you can't put measurement to)*

The radical change model. Major redesign or terminate and replace the current system.

The no-decision model. Is indecisive or consciously decides to do nothing

---

the basic policymaking model, or model mixes, in use by the system decisionmaker(s) leads the analyst to policy strategies with which the decisionmakers are most comfortable and also to the possibility of offering alternative policymaking models to the decisionmaker as well as alternative policies for specific problems.

VALUES ANALYSIS *(must know when, and where to place your values)*

Values are things or principles preferred. The values of an individual, group, or society are standards of desirability and evaluation independent of specific situations. They are what humans want and feel to be the reason for existence. Values are linked to tacit knowledge and influence beliefs, rationale for choice, life styles, and commitments. Values regulate the political process and the managerial process and lie at the heart of resource allocations. They are the lenses and filters through which the world is viewed. Values are mainly extrarational and exist because of complicated historical, geographic, cultural, psychological and socioeconomic variables rather than as a result of any rational decision process, although rational thought, time passage and environment may alter values. The integrity and quality of a system is often measured in terms of a set of values. For these reasons values analysis is essential in systems analysis and policy sciences and becomes one of the three fundamental categories of research methodologies described in Chapter 6.*

Values sources are roots (family, environment, schools, peer groups), culture, professional commitments, and social affiliations. For values analysis the indicators of values are those sources plus system choices, priorities in resource allocation, rewards and punishments, goals, system attention, leadership and management style, asserted policies, tacit policies (which must be deduced but are often more accurate for values analysis than asserted policies), advertising, observed biases, editorials, and position papers.

Values tend to be concealed by their extrarational nature, by taboos, and fears, and by defenses arising from repressions. They are commonly measured in absolute terms (i.e., "good" or "bad") using political, moral, ethical, cultural, or artistic standards of evaluation. To be more useful in analysis our methodologies must lead us to more precise identification and understanding of system values.

The best methodologies available are behavioral research of the indicators mentioned above, analysis of social-political ideologies as normative values systems, decision analysis (to include sensitivity to alternative values), budget analysis (resource allocation being one of the best indicators of values), interview, survey and Delphi, and identification of the meanings and dimensions of basic values of target individuals or groups.

The specific areas of concentration in values analysis are values implications of policy in terms of opportunity costs, values consistency in the system, absolute values (nonnegotiable) versus trade-off values, limits of feasibility for values explication, values mixes, values conflicts, values reinforcement or change, and values content of future assumptions.

For cross-cultural analysis values offer an extra media for interpreting, understanding, and predicting, which is often more reliable than rational analytical findings. For example in attempting to describe nations as developing, semideveloped, or developed the assumption that "a developed nation is one with a history and future high probability of a stable social and political set of values and institutions on which economic and social progress can be legitimized" can lead to additional insights to balance economic based definitions.

Values analysis should be a component of any normative research (i.e., investigating "what should be"), policy analysis, system design, redesign, or improvement, political feasibility determination, or future study. The systems analyst must recognize, however, that there are definite limits to the degree to which a set of

---

*I wish to make one distinction at this point. I use the word "values" for the meanings above. I will use the word "value" in the utility sense of economics and engineering. I find that this distinction can avoid the confusion that inevitably results when the word "value" is used to mean both preferences and utilities. There is only one familiar term in systems analysis where my usage cannot apply—that is "value judgment" which means a judgment made from a values set.

system values can and should be explicated. Explication of values held implicit due to their potential negative effects on consensus, leadership authority, or coalition maintenance can be traumatic or even destructive to the system. Herman Kahn used to illustrate that phenomenon at Hudson Institute Seminars with a story using a family metaphor. Paraphrased, the story goes as follows; picture you and your spouse at the dinner table with your two children and you decide that should a fire occur in the home and you have time to save only one child you will save Jim, not Irene. Furthermore you tell the children that decision. That is an example of implicit values that should remain implicit for as soon as they are made explicit (if, indeed such were possible) it becomes divisive or destructive to the family unit. One can easily extrapolate the example to companies, agencies, or nations in relation to managerial policy or political/social ideology.

Values analysis helps to answer the question for managers and decisionmakers of "Who are we?" If you have ever had the experience of attempting to obtain concurrence for plans from a Board of Directors when basic values were in conflict, and left implicit, you will recognize the value (in this case "utility") of accomplishing a values analysis to obtain first the consensus of the group on the questions of "Who are we?" and "What do we prefer?" This exercise may lead to the need for secondary agreement (agreeing on what cannot be agreed upon but acting on those areas of agreement) or even tertiary agreement (agreeing that agreement is not possible). If tertiary agreement is necessary then changes in the system or in the composition of the Board will have to be made before the values conflict can be resolved. Values analysis also helps to sort out management theory for practical application, helps to identify where individual values and system values diverge or conflict, can make the resource allocation job much easier unless unresolvable values conflicts are brought to the surface, and is an indispensable tool for personnel recruitment as we want people working in the system who have enough differing values to add creativity but not so different that disruptive conflict will result. Finally, values analysis puts logic into issue emergence. American deaths in the Vietnam War became a serious issue, whereas American deaths on the highways of America, occurring at several times the Vietnam rate, did not become a serious issue. Dying on the highway continues to be tacitly judged consistent with American values.[25] Dying in Vietnam was not.

EXTRARATIONAL VARIABLES CONSIDERATION

Table 1.2 outlines the theory of personal knowledge of Michael Polyani, which provides the concept of tacit knowledge to explain: (1) how we know more than we can tell; (2) why what we know and can tell is accepted as true; and (3) how a scientist can have a valid preknowledge of a problem that has not been completely explicated. Polyani's theory of personal knowledge solved the 2000-year-old paradox of Plato's *Meno* wherein Plato stated that if all knowledge is explicit then search for the solution to a problem is an absurdity for either you know what you are looking for, and then there is no problem, or you do not know what you are looking for and cannot expect to find anything. Plato's solution to the paradox was to quote Socrates as stating that all discovery is a remembering of past lives. Tacit knowledge is a much more satifying and scientifically useful explanation.[26] Alfred North Whitehead believed that the importance of philosophy lies in its sustained effort to make tacit theories and knowledge explicit and thereby capable of criticism and improvement.[27] The process of intuition can be viewed as a manifestation of tacit knowledge, however, psychological theory explains intuition from a different scientific paradigm.[28]

Table 5.3 provides a list of extrarational processes of the human mind that directly influence behavior and decisions without an intervening reasoning process. One of the early philosophical assumptions of systems analysis was that human reasoning powers will provide superior results to decisions flowing from extrarational processes. My position is that systems analyses considering extrarational processes as a part of their rational analysis will, on the average, produce superior

Table 5.3 *Extrarational Processes**

Process	Definition
Judgment	Decision wisdom stemming from experience
Intuition	Knowledge discerned by the mind without reasoning
Creativity	The mental process that brings into being new patterns, configurations, and relationships
Serendipity	Chance discovery through nonrational means
Tacit knowledge	Knowledge learned through living (Table 1.2)
Faith	Acceptance of truth not certified by reason
Love, joy, sorrow, hate, fear	Affective states of consciousness stemming from emotions
Charisma	Ability to inspire loyalty and admiration
Loyalty	Allegiance and perceived obligations to people or groups
Perception	Ability to understand
Will	Determination to reach goals (the addition of patience and stoicism extends will over time)
Politics	Processes of who gets what, when, and how (portions rational)
Extrasensory communications	Communications outside the normal range of the senses
Prophecy	Ability to predict forthcoming events

*Those processes of the human mind that influence behavior and decisions without a rational reasoning process occurring or where the extrarational processes predominate over the rational ones. Extrarational processes are different from "irrational behavior" as, for instance, caused by the stress and anxiety of threatening situations, but there are linkages between extrarational processes and irrational acts.

results to analysis limited to either rational or extrarational components. I specify "on the average" because there is ample evidence to show that decisions stemming purely from the judgment of experienced leadership has produced high-quality policymaking and is still the norm for policymaking in a large percentage of systems around the world. The converse proposition—which is that high-quality policy-making for human systems can emerge from purely rational analysis—is less supportable.[29] A management science capable of guiding effectively complex human systems will, for the foreseeable future, involve a combination of the science of rational analysis and the art of extrarational analysis. Rational systems analysis can provide tools, it cannot provide wisdom.

POLITICAL VARIABLES

Perhaps the most difficult area to deal with in systems analysis is political feasibility, or the probability that any policy alternative will be accepted and implemented. One of the best known manifestations is giving the boss what he wants. The more powerful, authoritarian, and doctrinaire the leadership and the more insecure the policy advocate(s) the higher the probability that the boss will be given not only the policy decision he is known to prefer, but none other. This phenomenon combined with the normal conservatism of the incremental policy-making model can infuse a system with the concept of "the more feasible the better" and can be a serious constraint on the identification of innovative policy alternatives. Furthermore, not everything feasible is desirable, applicable, or effective for the problem at hand.

The concept of political feasibility embraces governmental politics and organizational politics, as well as client acceptance of the solution in addition to top management acceptance.* Political feasibility, political power, and consensus building needs operate at every level of organizations. Elsewhere I have described some of the dilemmas this produces in civil-military systems.[30] Political feasibility is not an evil to be avoided but rather a normal phenomenon of public and private life that needs to be studied by the analyst. Political feasibility is also the process through which leadership transmits its judgments and values concerning policy directions. Political feasibility is ephemeral with changes being difficult to predict. System crises usually change political feasibility. An almost universal side-effect of natural or human-produced catastrophe is that the political feasibility environment is suddenly changed. Examples run the gamut from feasibility of improved aircraft safety systems following a mid-air collision of civil aircraft to aggressive attack across national borders.

The problem for the systems analyst is to understand the political feasibility environment, but not be completely limited by it—a difficult assignment. Completely ignoring political feasibility is one of the best ways to have an analysis discredited by leadership. On the other hand failing to offer alternatives that may

---

*In organizational theory terms the human relations and psychosocial system developments are relevant here. See the Bibliographic Essay references for Chapter 5.

be outside the current domain of political feasibility, but nevertheless preferable, is an abdication of the objectivity expected of the analyst.

Where system absolute values are explicated, political feasibility analysis identifies those areas where leadership is known to be receptive and those other areas that are taboo. At the national level it is not unusual to have explicated goals that are strongly associated with political and social ideology of the leadership in power. The recommendation of an analyst that would run counter to those absolute values would not only have an extremely high probability of rejection but might also have serious detrimental future career implications for the analyst.

A legitimate question to ask is: "Is analyzing political feasibility politically feasible or even desirable, in a system?" The answer is that it may not be. This is one area where an outside analyst has both advantages and disadvantages. The outside consultant does not have the tacit knowledge of the insider concerning the political feasibility environment. On the other hand, the outsider is not constrained by political feasibility to the same degree as the insider as his social and professional position is only indirectly, or perhaps not at all, hinged to the leadership to which the policy alternatives are being proposed. However, one result of failure or inability to analyze political feasibility will be that prediction and contingency studies may be irrelevant for policymaking and force decisions devoid of formal consideration of what is or could be politically feasible and perhaps essential for the system.

As the next chapter will point out, the Systems Analysts must do their own formal consideration of political feasibility variables. Whether or not all of that portion of the analysis is passed on to the decisionmaker is a value judgment the analyst must make after gaining knowledge of the contextual environment of the system. There are no firm guidelines to wend ones way successfully through the political feasibility pitfalls.

CROSS-CULTURAL VARIABLES

Cross-cultural variables are related to extrarational ones in that they are largely subjective, nonquantifiable considerations for the systems analyst. Table 5.4 provides some of the management oriented cross-cultural variables to which analysis results can be extremely sensitive. Conversely, if neglected they can produce distortions or even the "Butch" of Herman Kahn—a completely mistaken assumed notion or fact used as a basis for analysis.[31] The modern classic instance of cross-cultural communication Butch occurred in Warsaw moments after President Carter arrived December 29, 1977 for the first visit to Poland of an American President. His statement to the welcoming dignitaries and citizens that "I have a desire for peace . . ." was translated into Polish to mean: "I lust for Poles." Butches like that can be rectified with rapid diplomatic action. Butches in systems analysis are often not as rapidly identified but can be just as painful later. The interesting thing about the variables in Table 5.4 is that they have equally relevant applications

Table 5.4 *Cross-Cultural Variables in Systems Analysis*

The following variables should always be considered in cross-cultural analysis and can be significant for intra-cultural analysis as well:

Management theory often incompatible with other cultures and values

World views toward: societal direction; progress; growth; time; decisionmaking; efficiency; change; planning; egalitarianism; materialism

Environmental variables: national consensus and pride; political stability and constraints; international economic linkages; education and literacy; historical, social, and religious values; language; rates of social change

Organizational variables: authority and leadership (social status versus merit); centralized versus delegated; paternalistic versus impersonal; managerial entrepreneurship; ownership (family, partnership, government, stockholders); employee morale, job satisfaction, work-orientation, age, sex, loyalty, salary, benefits, job security, mobility, creativity, face-saving needs, acceptance of responsibility; organizational learning and use of data; risk taking

Cross-cultural sensitivity and empathy of the systems analyst

---

to what might be assumed as a homogenous managerial entity like IBM or a national ministry of defense. It is quite possible that differing world views will exist between individual decisionmakers at differing levels or geographic locations of the system.

FUTURE STUDIES

Scientists used to take the position that they should not be concerned with what ought to be thus tacitly accepting social and political values through the failure to critically examine them or consider them relevant. As recently as the decade following World War II it was considered naive to study the future. Today, managers, leaders, and politicians are considered naive or incompetent if they are not identified with future studies. Future studies, themselves, are not new, however. Forecasting from the movement of stars, from tea leaf patterns, or the entrails of birds has occurred since prehistory and theology provides a rich source of normative futures. Actually, most systems analysis is research for improved systems in the future. Even analysis of historical failures of systems to adapt have a motivation of producing knowledge applicable for today's and tomorrow's systems. Nevertheless, future studies are a unique analytical orientation in themselves and all good systems analysis must include some consideration of the future.

H. G. Wells' was the first author to popularize studies of the future with his book *Anticipations* published in 1900 and was also the first to consider technology as an independent variable for the future. The 40 years between the 1900s and the

1940s saw the emergence of science fiction with its fantasies of the future and some romantic and metaphorical literature but future studies lacked methodology, knowledge of the social systems, or values sensitivity. The 1950s can be considered the beginnings of a Western rational future studies discipline as the United States Department of Defense and the first think tank research organizations employed operations research specialists, systems engineers, and cost-effectiveness analysts. Bertrand de Jouvenal wrote the first theoretical and conceptual study of the future with his 1964 book, *"L'Art de Conjecture."* Serious scientific oriented investigation of the future, then, is about 25 years old. In that quarter century new national and international organizations to study the future have sprung up around the world and on university campuses. Why now?

Prior to the 20th century future studies tried to predict, to understand or to theorize so that the future could be anticipated and endured. It wasn't until the 20th century that the need to control human destiny, not merely endure it, became evident. Governments began to deliberately intervene in more areas of life for specified ends. The scientific, industrial, and technological revolutions made the world smaller, more interdependent with nations more vulnerable to the effects of other nations' policies.

The dangerous impacts of potential failures in the physical and social worlds increased exponentially. Some of the failures would be irreversible such as failure to prevent strategic nuclear warfare, failure to match population growth with food and resources, failure to preserve the earth's biosphere or failure to safeguard against biological catastrophe through mistakes or malevolent design in genetic engineering. There was evident need for long-range planning to rebuild cities, eradicate disease, reduce population growth rates, respond to rising expectations of citizens, and to develop the new nations gaining independence following the colonial era. This led to a general acceptance of what Geoffrey Vickers has called "The end of Free Fall" where active intervention to influence human destiny for the better must replace passive acceptance of fate.[32] The concept underlies the syllogism at the beginning of this chapter describing the motivations of policy scientists. An appreciation set for that fact is an essential conceptual tool for the Systems Analyst.

A set of guidelines and a few methodologies for the future studies are provided in Table 5.5. In spite of the fact that one may get approval in the abstract that future studies must be a major activity for organizational and societal direction systems there are a number of organizational barriers to future thinking and future studies that the analyst must confront. First is that current problems and operations normally preclude time for future planning. Furthermore, in an era of rapidly rising personnel costs top management will use their best people for line operations, for crisis management problems, or will reduce the planning staff capable of accomplishing future studies. Second, most leadership prefers short-range concrete projections and specific objectives as opposed to considerations of uncertain futures or utilizing todays resources to create future capabilities for as yet unspecified requirements.

Table 5.5 *Guidelines and Methodologies for Future Studies*

Guidelines—Future studies should*: Create and/or identify a spectrum of imaginative alternative contexts, futures and policy requirements; explicate assumptions and values internal to the study and projected to the future; relate alternative futures to present policymaking for sensitivity analyses; consider wide ranges in political, technological, and economic feasibilities; identify important policy issues and potential crises for the future from past and current trends including those that will have no precedent; suggest structural, procedural and personnel characteristics necessary for the policymaking system to utilize future studies; employ methodologies permitting concise and understandable documentation and presentations to policymakers

Methodologies for future studies: Trend extrapolation; use of analytic models, metaphors, scenarios, and historical analogies; technological forecasting; political forecasting; systems dynamics and cross impact analysis (e.g., interrelationships between population growth and development); Delphi techniques; outcome arrays; creativity-enhancing techniques

---

*Discussion of many of these guidelines can be found in Herman Kahn and Anthony Wiener, *The Year 2000*, 1967 and Yehezkel Dror, *Ventures in Policy Sciences,* 1971.

Reasons for this preference are linked to some well-known decisionmaker traits. One is simply a low tolerance for ambiguity or risk by public or private decisionmakers. Another trait is more associated with public policymaking and the political system cycle for electing politicians in a democracy. Representatives elected for short periods can do themselves political damage by committing resources of today too far in the future. It mortgages future power, which in turn reduces freedom of action and political leverage. Finally, having specific goals and objectives to meet in the future are formal statements by which leadership will be judged.* Therefore, future goals tend to be incremental ones reflecting low-risk strategies. There can be exceptions to this barrier when a strong leadership with a broad popular consensus can rally support for ambitious, nonincremental goals, but the exceptions merely help to identify incrementalism as a prevailing mode.

Still another barrier to future studies is organizational resistance to power shifts. Political power is relative and can be a zero sum game—that is leader "A" loses when leader "B" gains. Future planning gives power to those who plan or gives leverage to those whose recommendations are accepted. It was the resistance to power shifts caused by the introduction of systems analysis into the U.S. Government in the 1960s that was the major stimulus for the Congressional Investigation of systems analysis and PPBS and the subsequent 1970s reduction of influence

---

*This phenomenon applies to systems analysts and consultants as well. The difference is that decisionmakers bear the responsibility for accepting the advice of the analyst. Part of the analysts' job is to include a risk assessment for each alternative policy identified.

of systems analysts in the government.* A significant contribution was the oversell by bright and enthusiastic analysts of the capabilities of systems analysis and PPBS, which produced results justifying criticism. Connections between future studies units and policymaking units in systems for the above reasons, tend to be weak. As a result, it is difficult to follow the guidelines for future studies in Table 5.5. Future studies often are accomplished by the wrong people with inadequate resources and their recommendations emerge as not credible or feasible. This does not negate the need for future studies. It does mean that analysts have a difficult job accomplishing good future studies and decisionmakers have an equally difficult task in evaluating the future studies they receive.

CHAPTER SUMMARY: SYSTEMS ANALYSIS
AND POLICY SCIENCES MIXED

Expanding the economic and mathematics based systems analysis of the 1940s to 1970s with the qualitative macro and social orientation of policy sciences makes a quantum jump in complexity for the analyst but interjects a corresponding amount of realism. The analysis becomes more difficult, less precise, but much more relevant to the real social and political world of human systems. This, in turn, makes policy recommendations and alternatives more feasible and acceptable to decisionmakers and clients, as well as more effective in improving or designing systems.

---

*Political concern over new managerial concepts that impinge on the American social and political fiber is normal. The scientific management concepts of Frederick Taylor and his colleagues produced congressional hearings in 1912. The Planning Programming and Budgeting System (PPBS) brought to Washington by Secretary of Defense Robert McNamara in 1961 drew Congressional fire in 1966. The Zero-Base Budgeting system of resource management directed for all U.S. Departments and Agencies January 1, 1978 has already generated debate at the political level.

6

Research Methodology
for Systems Analysis

This chapter and the two following form the methodological heart of the book. Methodological steps for systems analysis are provided which include: fundamental categories of research (behavioral, values, and normative); feasibility (economic, technological, and political); the relationships of those methodological components to problem solving in human systems and finally the method of communicating results of the analysis—the Systems Analysis Briefing which has immediate and long-range implications for the analyst, the decisionmaker, and for future development of democratic societies.

There are three fundamental interrelated categories of research methodologies for systems analysis: (1) behavioral research; (2) values research; and (3) normative research. Behavioral research asks the question: "What is?"; values research investigates: "What is preferred?"; and normative research asks "What should be?". The three feasibilities—economic, technological, and political—must also be met by a systems analysis. The quality of resulting systems improvements and design will be positively correlated to the quality of research accomplished.

BEHAVIORAL RESEARCH

Behavioral research seeks to answer the questions of: What? When? How much? and How many?. The assumption is that things can be seen as they are and that reality can be discovered. Behavioral research makes descriptive statements about things, events, relationships, and interactions. It observes. It counts and measures. Scientific claims to knowledge are usually made on the basis of behavioral research. It uses both inductive and deductive reasoning but leans toward the inductive as a preferred mode. Values are outside the scope of pure behavioral research—or, at least an attempt is made to distinguish facts from values and to then deal with the facts within the larger values set. Because facts and values are never really separated in human systems does not reduce the usefulness of the behavioral research model, but it does alert us to its inter-relationships with the other two research categories. Validation of behavioral research—an essential component of the scientific method—occurs through repetitive observations and comparisons. The fundamental scientific statement of behavioral research is:

IF CERTAIN FACTS ARE OBSERVED OVER TIME THEN A KNOWN RESULT WILL OCCUR WITH STATED PROBABILITY

Behavioral research is the bread and butter data discovery mode of science. It is best exemplified in the physical sciences where experiments produce theories with long lives capable of continual validation. For an example from chemistry consider: "if two atoms of hydrogen are mixed with one atom of oxygen under appropriate conditions of temperature and pressure then the compound H_2O (water) will result with a stated probability (extremely high)." That one was easy, as is the following engineering sample: "if specified material strength standards are used for a bridge construction then known stress and pressure limits can be reached before structural failure will occur with a stated probability (also extremely high but perhaps lower than the chemistry example)." Inherent in computed probability will be an estimate of the degree of confidence held in that probability developed from behavioral research and statistical data accomplished over time.

For an education example consider the statement: "if phonetic methods of reading are substituted for rote memory methods in elementary schools, within prescribed conditions, then a reading improvement will result for those students (compared with a control group continuing with the rote system) by high school graduation with a stated probability." Having switched to a subject with a higher human component it is to be expected that our stated probability and confidence will be lower and that the research will be more difficult. For a much more complex, but still scientific, statement consider: "if nuclear superiority/equality/ sufficiency (you pick the one you prefer) is maintained by nation 'X' in comparison to nation 'Y' then nation 'X' will deter nation "Y" from military aggressive action against nation "X," or its allies, with a stated probability." This one is much more difficult to validate and even controversial but a current real-world behavioral statement that has high significance to decisionmaking for defense and resource allocation in nations possessing nuclear weapons, or those considering such development.

My final example of a behavioral research statement deals with the subject of this book: "if systems analysis is accomplished, within parameters prescribed herein, then systems improvement will result with a medium to high probability." Since the systems analysis described has been in use for some time—although perhaps not conceptualized in the same manner—it is not necessary to wait for feedback from application of the tools and techniques to completely validate that statement. However, to survive as a useful scientific statement it must continue to be validated and the anomalies must be investigated. Both of those efforts—continual validation and aggregation of anomalies—are the job of behavioral research into the effectiveness of systems analysis as a meta problem solving tool.

My point is not only that the behavioral research fundamental statement has applicability throughout the complexity range of problems but that it is extremely important to keep the behavioral questions conceptually separated from the values and normative questions of the other two categories for initial analysis purposes. Conclusions and recommendations for solutions to real world problems must flow from a combination of the results of research into "What is," "What is preferred," and "What should be?" but mixing those elements too early in the analysis is a common mistake.

Before going on to those other two categories, however, I want to point out the three places in behavioral research where, in spite of attempts otherwise, values and value judgments enter. The first instance is in the definition of the boundary conditions of the system and the problem of interest to the analyst. System definition reflects the preferences (values) of people and groups, and the values inherent in issue emergence have correlation with what those systems identify as problems. Secondly, in the selection of facts or events to observe—and methodologies picked to observe them—behavioral research signals its values as every such selection implies many explicit and implicit rejections. Thirdly, values have a great deal to do with the overall assumption of behavioral research of its own objectivity. Facts tend to be qualified by the values set in human systems as well as by the values set of those doing the analysis.

VALUES RESEARCH

Values are things or principles preferred. Values sources, indicators, characteristics, values analysis methodologies, and some specific focus areas for values analysis were covered in the Chapter 5 discussion of policy sciences and policy analysis. The research of values constitutes one of the three separate categories of systems analysis research. Values research provides the analysis to justify that the end is worth doing, the means are acceptable, and the resulting improvements to systems are "good." Values research seeks to answer the questions of: "Why?, For what?, For whom?, With what commitment?, With what risk?, and With what priorities?" Values research addresses the values issues directly through values identification and analysis. The assumption of values research is that values are a major determinant of action and behavior in human systems. The fundamental statements of values research are:

THE SYSTEM PREFERS . . .

SELECTED COMPONENTS OF THE SYSTEM PREFER . . .

Television viewer surveys can provide numbers of people watching individual shows or average number of hours people watch TV per day. That is behavioral research. Values research can identify the preferences and the qualitative standards exhibited. The last category, normative research, provides analysis on which to prescribe what programming people should watch or what programming television producers should provide the people (depending upon one's preference for the role of television in society).

NORMATIVE RESEARCH

Normative research seeks to confirm the assertion or hypothesis of what "should be" by identifying and validating actions and means to achieve those prescribed ends. As normative research works with statements of what should be done, it has an inherently idealistic component. It uses predominantly deductive reasoning from abstract generalizations to conclusions about specific action or events. Future studies are an essential part of normative research. It is in the normative research category that alternative solutions are identified or created and where the linking of research to feasibilities occurs. Validation of the ends prescribed can only occur in the future through behavioral research or social reactions to what actually occurs. Validation of the means identified by the normative research to achieve those ends can be incrementally validated through indicators for applicability, feasibility, desirability, and effectiveness at periodic intervals. The fundamental normative research statement is:

IF YOU WANT CERTAIN RESULTS, THEN DO PRESCRIBED
ACTIONS, WITHIN SPECIFIED CONDITIONS,
AND YOU WILL SUCCEED, WITH STATED PROBABILITY

In normative research values interface at three places: first in wanting the results because they are preferred (i.e., "good"); second in selection of means considered desirable, feasible, applicable, and effective; and third, in the assumption that people and institutions impacted by the policy in the future will have no values conflict with conditions resulting.

In time-line terms, behavioral research is involved with "what is" for the past and present; values research covers the entire past-present-future time spectrum; and normative research is involved with future expectations (see Table 6.1). The three research categories cannot, and should not, remain isolated from each other throughout the research. Reality in complex human systems combines individual and organizational preferences, observations of what exists and prescriptions for what should occur in the future. Systems analysis must synthesize these areas also—but only after attempting to sort them out individually first. A common phenomenon in evaluating, judging, and analyzing the worth of human systems is to inextricably confuse the reality of what exists with strong—often emotional—

Table 6.1 *Systems Analysis Research, Problems, and Solutions*

Research Categories	Time Line		
	Past	Present	Future
Behavioral	← ———————————→		
Values	← ——————————————————————→		
Normative			← —————→

If: The actual system state (behavioral + values) does not equal the desired system state (values + normative)

Then: A problem exists and the alternative solutions identified in normative research must meet the criteria of: *economic feasibility; technological feasibility;* and *political feasibility*

feelings of what should be, which are heavily biased by values. This is not a recommendation for valueless analysis. On the contrary, it is an appeal and methodology for understanding our system values, their impacts, and their linkages to the rest of our analysis, and to the success or failure of the enterprise.

WHAT IS A PROBLEM?

Those three research categories become a vehicle for the analyst to more precisely identify the nature of the problem. Table 6.1 diagrams the relationships in terms of problem definition. A problem exists any time the actual state and the desired state are not equal.[33] This can occur when (1) behavioral research indicates changes have occurred (e.g., "profitability indicators have dropped"); (2) values research reveals a values dissonance (e.g., "new management holds values that are reordering priorities between productivity and employee satisfaction"); (3) when normative research calls for changes (e.g., "the company should decentralize its operations and increase its marketing efforts"); or (4) any combination of events and analysis that indicates a deviation between actual and desired system states.

With that definition of a problem, is it likely that problems in human systems will ever disappear? Very unlikely. That surfaces an important point for systems analysis—the need for analysis is not periodic, but continual and iterative. The management function of control within systems should be conceptualized to provide continual or frequent monitoring of indicators and evaluation of those indicators.

FEASIBILITY

The problem solution must meet the three feasibilities—economic, technological, and political. Political feasibility, or the probability that the policy alternative will be acceptable to the decisionmaker(s) or clients was covered in Chapter 5. Economic feasibility is the probability that resources will be available. Technological feasibility is the probability that the technological and scientific goals for the system will be met. Although the data used to determine the three feasibilities comes from different sources, the three feasibilities are not mutually independent. They are mutually supportive. All are necessary and no two alone are sufficient for a solution to emerge. Economic feasibility has a great deal to do with the status of technology. Technological feasibility may well depend upon the size of the budget—an obvious economic input. Political feasibility may be reversed from a "no" to a "yes" through the accomplishment of professionally done economic and/or technological feasibility studies. Due to this interdependence phenomenon, the decisionmaker-analyst team must consider priorities, constraints, potential benefits, and pitfalls when structuring research effort into the three feasibility areas. This identifies another art-science mixture for systems analysis.

There are other questions to ask of both the methodology utilized and the results obtained. Feasibility inquiry answers the question of "can we?", but all actions that are feasible are not desirable. Other useful questions are: Is it applicable (right solution for the right problem)? . . . Are the analytical tools necessary and sufficient? . . . How sensitive are findings to methodology used or to hidden bias (values)? . . . Will recommendations be effective (i.e., meet goals within values parameters)? . . . Will they be efficient (revealed by cost-benefit and opportunity cost analysis)? . . . Are means and ends ethical and moral (these "should we?" kind of questions are a part of the normative research)?

Problem statements are sometimes put in the form of hypotheses that are tentative statements of the expected relationship between two or more variables. Hypotheses can be in a declarative form (e.g., "saccharin produces cancer"), a positive or negative correlation form (e.g., "there is no relationship between the use of saccharin and cancer"); a question form (e.g., "does the use of saccharin produce cancer?"); or in a null hypothesis form (e.g., "everything causes cancer"). These are specific scientific problem statements, which are quite different from an objective of "improving system quality." The systems analyst may deal with several forms of problem statements or questions on a given analysis (e.g., "we want to improve organizational effectiveness and will investigate the hypotheses that participative decisionmaking will do so"). Hypotheses perform some of the same functions as models, but on a more limited scale. They restrict and limit the research topic to statements of expectations that can either be supported or rejected; they reduce expenditures of time and effort; and they help to explain relationships between variables.

The methodological job of the systems analyst, then, can be conceptualized as accomplishing research within, and between, the three categories of behavioral, values, and normative research and seeking alternative courses of action that meet the three feasibilities. Whether or not the analyst proposes one preferred solution or several alternatives will depend upon the relationship of the analyst to the decisionmaker and the nature of the problem. In some cases the analyst may be the decisionmaker. In the majority of cases where the analyst is advising the decisionmaker the final methodological step for the analyst is to put the results of the research and analysis in a form understandable to the decisionmaker.

THE SYSTEMS ANALYSIS BRIEFING

All is lost unless effective communication can occur between the analyst and the decisionmaker.[34] In Table 1.3 I have shown "Communications of Findings to Decisionmaker" to be a major step in the scientific method in systems analysis. Beautifully prepared, lengthy, and detailed reports are fine and in most cases need to be prepared to provide adequate documentation and description of the research effort. Few top executives have time to read them. It is true that top executives in private and public systems have staff officers who can read reports in detail but they will rarely do so without direction from their leader as most human systems executives are quite capable of keeping their staff occupied. Furthermore the dangers of a staff officer condensing your study in unfortunate directions is a real possibility. So initially it's the traditional problem of "first you have to get his attention."

Let's assume that you have surmounted the politics of that situation—that is, you have obtained 20 minutes of the boss's time. Twenty minutes? Yes! That's not an unreasonably short time to give the boss an appreciation set for the problem scope, methodology, findings, alternative policies, and impacts and your recommendations for action or further study. Obviously 20 minutes is an arbitrary time period. The boss may want to spend all day tracking the analysis. Even so, the advantages of being able to summarize the study in such a short time remain. Table 6.2 provides the guidelines, advantages, and pitfalls of the systems analysis briefing. No two briefing situations are alike as no two human systems research and analysis problems are exactly alike. Furthermore, there is a whole spectrum of possible relationships existing between the analyst and the decisionmaker. If the decisionmaker contracted for the study, he or she may have been actively involved through its phases, have read your executive summaries, your periodic progress reports, or even carefully read the entire written report. On the other hand, your briefing may be coming as a new specific subject, but one obviously within the decisionmaker's sphere of responsibility and hopefully also the decisionmaker's competence.

The systems analysis briefing may, therefore, consist of notes scratched out on the back of this morning's hotel breakfast check* or a computer generated visual probabalistic decision display as described in Part II, Chapter 18. My own prefer-

*You may prefer the Abe Lincoln "Gettysburg Metaphor" of "back of an envelope." There could be some psychological advantages to using the envelope particularly if you match the envelope with the occasion.

Table 6.2 *The Systems Analysis Briefing: Guidelines, Advantages and Pitfalls*

Guidelines: THE GOAL of a systems analysis briefing is to demonstrate to the decisionmaker that the findings of your research flow from a structured, rational, scientific analysis that has also addressed extrarational considerations, and that the recommendations, if implemented, will—within acceptable risk and values parameters—result in net benefits

The Briefing Should: present problems, scope, methodology, findings, clear alternatives with associated impacts, conclusions, and recommendations for action or further study; clarify concepts and relationships between variables; breakdown the problem into subissues susceptible to decisions; test conclusions for sensitivity to assumptions, values, and uncertainties relating to alternative recommendations and alternative clients; identify main interconnections with other issues and systems; be designed for the audience; be prepared to counter the criticism of the strongest antagonist; have a creative component; utilize visual aids; be short

Advantages: Condenses and capsulates concepts and relationships in a form consistent with the decisionmaking process and subject; a time saver; . . . a discussion and decision focal point; educates; . . . can be a consensus building device . . .

Pitfalls: Too concise; if poorly done may result in premature rejection of good alternatives; often used as an advocacy medium rather than a systems analysis (i.e., more politics than science)

---

ences tend toward *some* preprepared audio-visual aids—flip charts (adapting size to audience), overhead viewgraph projections (similar to the viewgraphs utilized here as charts and tables), which have the advantages of rapid production and change capability and lend themselves also to making hard copies for distribution at the briefing, 35 mm slides, television playback equipment, or combinations of these media. Learning studies indicate higher comprehension occurs with combined verbal and visual communications.

The major goal of a systems analysis briefing is to demonstrate to the decisionmaker that the findings of your research flow from a structured, rational, scientific analysis that has also addressed extrarational considerations and that the recommendations if implemented will—within acceptable risks and values parameters—result in net benefits. Various briefing concepts were developed separately within government, the military, and business. Efforts to conserve time for busy leadership was the major motivation but other spinoff benefits emerged as the technique took hold—particularly in building consensus for controversial proposals, informing top management of ongoing research and activities, public relations, and as the focal point for critical decisions. The technique is now highly developed and widely used often with elaborately prepared and complex visual aids. The advantage that overshadows all others is that of forcing the analyst to condense, capsulate, and clarify ideas and to strip the mass of supporting data away leaving the meaningful variables and relationships, findings, and conclusions. If you have difficulty in extracting the essential thrust, objectives, and outcomes of the study into a 20 minute presentation you probably have been too absorbed with the many details, dimensions, and complexities and spending too little time on conceptualizing the effort and data into a coherent form that lends itself to decisionmaking. A briefing is a good form of mental discipline for the analyst as well as a time saver and facilitator for the decisionmaker.

The pitfalls of briefings are shown in Table 6.2. It is important to make a distinction between "advocacy briefings" and "systems analysis briefings." The former has political gain (who gets what, when, how?)[35] as a primary motivation; the latter is an attempt at presenting scientific knowledge and real alternatives for system improvement or design. Briefings tend to be used more as a device to sell a point of view or a decision already made than to educate for a rational decision although those two approaches are not always mutually exclusive.

Systems analysis and policy sciences have important roles to play in the transition of briefings from simple partisan advocacy toward education in the private and public sectors for more sophisticated handling of complex issues. When used as an advocacy briefing the overriding pitfall is that the decisionmaker, or the public, may be led down the primrose path to a sudden precipice rather than to the advertised little white cottage—particularly if the briefer is influential and leaves major portions of the values analysis implicit. Other than education of the briefing participants there are four possible decision outcomes to a systems analysis briefing. The decisionmaker can (1) say "yes" and direct the implementation of the study recommendations; (2) say "yes" to a portion of the recommendations and reject the others; (3) state an interest but ask for more analysis; or (4) say "no."

BROADER IMPLICATIONS OF THE SYSTEMS ANALYSIS BRIEFING

The increasing complexity of human system issues requires more sophisticated means for bringing the results of good analysis to the public as well as to the relevant decisionmaker. Increasing the policy contribution capacity of citizens, employees, managers, decisionmakers, and politicians should be a high-priority goal for any democratic nation. Democratic theory assumes rational choices by an informed electorate. Improved modalities for bringing systems analysis briefings to a wider range of people and groups, to school rooms and board rooms, and to government agencies at all levels are needed. The issue of whether or not television, radio, and news journal presentations—and especially editorials—are advocacy briefings, systems analysis briefings, or some mix of those extremes is relevant to these broader implications, and to the improvement of social and political processes. For instance, the first politician who successfully implements a television media system to provide constituencies with good systems analysis briefings and a Delphi method to collect individual citizen views on the issues will raise the probabilities of their successfully representing their constituencies, will provide a valuable educational media for current issues, make the democratic process meaningful to many who now feel isolated by the political system, and no doubt raise his or her own probabilities of reelection.[36]

SUMMARY

The methodological steps in systems analysis are summarized in Table 6.3 and illustrated with the real world case studies in Part II. As with any logical and linear exposition of a complex process it is a simplified model of the reality of systems analysis and is primarily heuristic—that is, intended to provide the analyst with a checklist of useful methodologies. The scientific method dictates investigation as free as possible from preconceived notions, biases, and culturally based "common sense" judgments, which if acted upon alone, can often lead to counter-intuitive results. Systems analysis conceived in a policy sciences framework provides methods to better analyze the qualitative variables along with the quantitative ones. There is no neat packaged approach to the solution of difficult problems in human systems. The analyst continually works the problem with the basic methodological tools available and builds new tools or acknowledges analysis limitations when those tools prove inadequate. Knowledge evolution will sharpen those tools.

Table 6.3 *Methodological Steps in Systems Analysis (A Combination of the Scientific Method with Systems Concepts)*

---

1. Explicate the problem, opportunity, or insight through breakdown of the subject into analyzable interrelated subcomponents and potentially rewarding directions for research

2. Accomplish behavioral research (what is) to define and describe the system and its subcomponents and environment

3. Accomplish values research (what is preferred)

4. Perform a systems analysis evaluation (two-step process to determine and judge quality)

5. Accomplish normative research (what should be):
   A) Establish what could be (alternatives)
   B) Decide what should be (recommendations and implementation steps)
      (1) Justification based on steps 2, 3, and 4
      (2) Determine feasibility (economic, technological, and political)
      (3) Conceptualize verification and evaluation processes

6. The systems analysis briefing to communicate with decisionmaker(s)

7. Validate results during and after implementation (two-step process again)

8. Recommend policy revisions if required.

7

Evaluation and Measurement

An incorrectly evaluated system will move toward disorganization. The two-step process of system evaluation is described with an explanation of why net output cannot be used as a criterion and how secondary criteria are selected to determine the quality of the system. Comparative standards and forms of measurement for judging the worth of the quality determined are provided with a corporate example. A principle of system quality leverage and an evaluation exercise are offered. The Part II Case Study at Chapter 12 demonstrates this evaluation process at a Nuclear Power Plant.

EVALUATION

We evaluate to determine quality as measured against a standard. The methodology for evaluation of a system is conceptually very simple but operationally often extremely complex. The process, in theory, involves two steps:

STEP 1: DETERMINE THE QUALITY USING CRITERIA
Ask: "What quality exists?"

STEP 2: JUDGE THAT QUALITY BY COMPARISON WITH STANDARDS
Ask: "How good is that quality?"

STEP 1: DETERMINE THE QUALITY USING CRITERIA

We would like to determine the quality by using total net output of the system as our primary criterion. If a jet engine is our system of interest it is possible to do so by measuring the engine thrust in units of energy. The total net output can be expressed quantitatively in energy units translatable into movement of an aircraft over distances through time. But is thrust really the only, the total, output? How about noise and atmospheric pollution? How about consumption of nonrenewable fossil fuels?

The problem with using net output as a primary criterion for determining quality is that we cannot measure, conceptualize, or even identify the net output of a complex entity. We just saw that it was difficult, although perhaps possible, with something as simple as a jet engine. The complexities rise exponentially with personnel evaluation and evaluation of human systems.

Consider the problem of identifying the net output of a corporation to use as a primary criterion for determining quality. Net output could conceivably include a mix of goals, services, profit, productivity, growth, security, capital investment, research and development, personnel salaries and benefits, training, compliance with government regulations, safety records, community good will, market diversification or share, or stockholder satisfaction. The organizational effectiveness example, Table 7.5, illustrates one approach to the problem of corporate evaluation.

Even more difficult is the problem of obtaining consensus on the definition of net output of an educational system. Experts differ on whether it should be knowledge, skills, professional advancement, economic development, socialization of children into a culture, values reinforcement or molding, or as a societal change agent, or some mix of those outputs. Furthermore educational output occurs over long periods of time and in most cases could not be expressed in comparable units with educational inputs in order to compute a net output.

For a final, and even more abstract, example one can say that the net output of a military organization is national defense. But how useful is that as a primary criterion for evaluating the quality of an army, a nuclear strike force, or the appropriateness of the allocated share of the national budget for defense expenditures? Should the U.S. national defense effort be evaluated as "good" since U.S. territory has not sustained a direct military attack since December 7, 1941? An answer of either yes or no to that question would have little operational or planning meaning for the Department of Defense, although the argument one way or the other might be used by advocacy groups with nondiscriminating audiences.

Since we cannot adequately define or scientifically use net output as a primary criterion for determining the quality of a system in Step 1 we must, and do, use secondary criteria for that purpose.

Secondary Criteria

Analysts use selected subcomponents as secondary criteria for evaluation of a system because they are considered—for good reasons—to be positively correlated with and more measurable than system net output (or primary criterion). Even if we cannot identify the net output of a jet engine we can select "pounds of thrust" as a secondary criterion of output, which is considered for good reasons to be positively correlated with and more measurable than net output. One "good reason" is that we know pounds of thrust as a unit of measurement allows us to predict jet aircraft performance through such useful ratios as the thrust-to-weight ratio and the time-to-climb ratio. Furthermore, the quality of the system can be stated specifically and quantitatively (e.g., 20,000 pounds of thrust), which allows moving ahead to Step 2 of the evaluation process, the judging of that quality by asking: "Is that quality good enough to satisfy the agreed engine and aircraft operational standards?".

Secondary criteria can be identified as components of input, people, structure, process, or outputs for evaluation. Once selected, secondary criteria become independent variables (those that are causal, effect the change, and may be deliberately manipulated by the system or analyst) to shape the quality of the net output as the dependent variable and primary criterion, which, as I have shown, rarely lends itself to evaluation. Table 7.1 provides illustrative examples of secondary criteria grouped by major subcomponents of a system. Table 7.2 illustrates graphically the "principle of system quality leverage," which accounts for the fact that small improvement in one or more of the components (e.g., the policymaking system or, conversely, fixation on others (e.g., productivity) could have significant leverage effect but in opposite directions. This is why the selection and continual review of secondary criteria for evaluation is critical to the success of the enterprise. Poor evaluation methods may allow secondary criteria to assume the role of a primary criterion due to the necessity to evaluate on a subsystem level. If return on investment, over time, is allowed to become the primary criterion of a

Table 7.1 *Illustrative Examples of Secondary Criteria for Systems Evaluation**

Subcomponent of System	Secondary Criteria
Inputs	Personnel quality; knowledge; values; capital; technology; income; and equipment
People	Knowledge; judgment; loyalty; commitment; motivation; management style and effectiveness; communication skills; human relations skills; experience and achievements; formal education; training; ethics; morals; age; sex; and distribution of qualified personnel within the system
Structure	Main structural forms (vertical, horizontal, matrix, mixed); size; degree of centralization/decentralization; geographic dispersion; units in charge of: policymaking, control, quality assurance, long- and short-range planning, relevant R&D, marketing, and systems evaluation; units interacting with the environment; and power representation and distribution within structural units
Process	Policymaking; communications between major components (e.g., labor-management, top management-middle management); information flow; product flow; monitoring and control processes; learning from feedback and crossfeed; goal setting; crisis management behavior patterns; management processes; and evaluation process
Outputs	Productivity; client satisfaction; profits; growth; security; market share; employee satisfaction; employee performance; goods; services; safety; good will; quality of life; and progress

*Secondary criteria may be measured qualitatively, quantitatively, or a mixture of both.

Table 7.2 *Principle of System Quality Leverage*

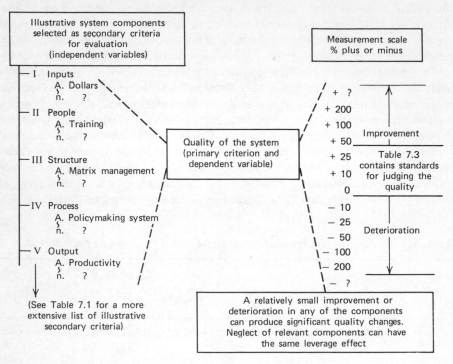

business organization as opposed to a broader criterion of organization health and productivity, a secondary-to-primary criterion transformation has taken place that could be costly, risky, or even fatal. This is the major utility of maintaining distinction between the primary criterion and the secondary criteria; both levels should be formally reviewed for adequacy on a periodic basis. If a system component is incorrectly selected for use as a secondary criterion, or assumed to be a relevant secondary criterion for the future because it was in the past, without challenging that assumption, the evaluation process will give leadership incorrect signals or may give warning signals (e.g., "our profits are down" or "our clients are unhappy") but no valid clues as to diagnoses of the problem or necessary corrective actions to take. This also can account for the not uncommon phenomenon of management's intuitive feeling that "something is wrong" in spite of the system's evaluation indicators being positive.

This is why "evaluation" is included in the list in Table 7.1 as a potential secondary criterion as the evaluation process, itself, may be faulty. For example, "bodycount" and "land controlled" were used by the American Administration during the Vietnam War as secondary criteria for determining the quality of progress of the war effort. Since the North Vietnamese government evaluation process used quite different secondary criteria for determining the quality of progress (e.g., sacrifice over time; will; military pressure; and reliance on the American people's increasing opposition to the war), the U. S. evaluation process distorted the subject of its evaluation and gave faulty signals. This is an example of a "tautological evaluation process." A tautology is logic that is true primarily by virtue of its stated form. If only secondary criteria are selected that are (1) measurable, and (2) indicate progress in the direction we want, then our evaluation process will indeed tell us what we want to hear—but it will be a self-deluding tautology isolated from reality. In these cases, where the primary criterion is not apparent or is extremely vague (e.g., "maintain a free South Vietnam") the analyst may want to observe what secondary criteria are actually being used within the system for evaluation and observe the degree of correlation that exists between secondary and primary criterion. Leadership may or may not be receptive to a criticism of its selected secondary criteria or the utlity of measurements made from applying them. The previous discussion of political feasibility applies in such cases.

It behooves the analyst and the decisionmaker reviewing the analysis to remember that system components selected for evaluation are assumed to be the most important or useful independent variables to represent system quality. A value judgment is made in that selection and should be justified. Past usefulness of a secondary criterion is relevant but not sufficient to assume that it will be positively correlated to the primary criterion in the future—particularly where the system is in a rapidly changing environment.

It is the job of the analyst to propose the secondary criteria and to design the evaluation process and mechanisms required to monitor that process. It is the job of the decisionmaker to be cognizant of the implications that some are being selected and others are not—and for which "good reasons." Failure to do so will separate leadership from the realities of system quality. In complex human systems this leadership requirement demands increasing knowledge and wisdom. In this case wisdom is defined as the ability to utilize knowledge to improve human systems in the direction of a "good" values set.

Once leadership is satisfied that the selected secondary criteria are, in fact, positively correlated to the primary criterion the answer to the question of "What exists?", to accomplish Step 1 of the evaluation process, can be determined through the use of a number of scales for measurement. Those scales are summarized in the "measured in . . ." box in Table 7.3 on "System Evaluation" as being binary, absolute, ratio, statistical or internal ranking. For instance the quality of the output of a jet engine can be measured in absolute terms of "20,000 pounds of thrust." The quality of a production line can be measured with a ratio of "cost/units produced." The quality of formal education in personnel evaluation can be measured in binary terms of "no, the applicant does not have a masters degree"; or "yes, she has four years experience"; or, "she has an ability to communicate orally in a clear, concise, and confident manner." Note that in this first evaluation step we are determining quality only, not yet judging it in comparison with standards. Recognizing this distinction between determining quality and judging that quality is essential in evaluation.

STEP 2: JUDGE THAT QUALITY BY COMPARISON WITH STANDARDS

Once use of secondary criteria have allowed us to determine the quality of the system we can then ask: "How good is that quality?" in comparison with standards. Table 7.3 lists standards for judging quality divided into rational and extrarational categories. This distinction helps to avoid the most serious potential pitfall of systems analysis—that of selecting secondary criteria for evaluation because they are the only readily measurable ones and because resulting numbers are capable of insertion into mathematical or economic formulas or into computer programs. There will always be one or more applicable standards from each of the "rational" and "extrarational" categories. Extrarational standards have always been the dilemmas and responsibility of leadership. Systems analysis can assist in that dilemma as well as with the rational standards. Extrarational standards are often intuitively judged as "good" or "bad" but considering them as providing enforcing, conflicting or neutral values to the system is more analytically useful. See Table 5.3 for a listing of extrarational processes.

Using rational standards we ask, "Is it true?" For instance, is it true that the marketing efforts of electronics company "A" can be judged as being of higher quality in 1978 than in 1977 using the standard of past quality measured in sales? We can make an objective "yes" or "no" answer and either give the Vice President for Marketing a raise or reprimand. Using extrarational standards we measure against a values set—usually the systems' but sometimes our own (which presents a dilemma for the analyst)—and ask, "is it good?" A problem with much analysis is that the extrarational analysis is not done explicitly and therefore influences the rational standards covertly.[37] The familiar line that: "I don't know why it's that way, it's just our policy" reflects the truism that employees often cannot rationally duplicate the judging process that has occurred by leadership, which always includes combinations of extrarational and rational considerations.

The reader will find illustrative examples in the case studies in Part II of the system evaluation process outlined here. A minimum of concentration on Table 7.3 will stimulate many possibilities for the analyst in standards trade-offs, impacts of standards selection, and standards mixing. One such exercise is provided in Table 7.4 where you are asked to consider what alternative standards might be used if the client and/or the decisionmaker change.

Measurement

The standard(s) selected in Step 2 of the evaluation process for judging the quality determined in Step 1 will place constraints on the method of measurement as well as provide a linkage between the two steps. To return to the jet engine example we determined that thrust was one appropriate secondary criterion of interest and that our hypothetical jet engine was rated at 20,000 pound rating compared with one or more standards from the list in Table 7.3. Any one of or a combination of the standards listed could be used. The engine could be judged unsatisfactory

Table 7.3 *System Evaluation*

Step 1:   Determine the quality by using criteria; ask: "What quality exists?"

Step 2:   Judge that quality by comparison with standards; ask: "How good is that quality?"

Standards for judging the quality

Rational Standards
- Optimal quality
- Past quality
- Leadership standards
- Quality of similar systems
- Planned quality
- Satisfying quality
- Survival quality
- Quality demands of clients
- Government/legal standards
- Professional standards

Extrarational Standards
- Managerial standards
- Ethical standards
- Moral standards
- Political standards
- Cultural standards
- Loyalty standards
- Artistic standards

Mixes of Rational and Extrarational Standards (the norm)

Measured In:

Step 1: Determine quality

1. Binary terms (yes or no)
2. Absolute terms (50 units/)
3. Ratio terms (cost/unit; births/year)
4. Statistical terms (distributions, averages)
5. Interval ranking (MTBF)

Step 2: Judge in comparison with standards

1. Absolute terms (too good, OK, unsatisfactory)
2. Ordinal terms (1st, 2nd,. . . last)
3. Superior–inferior (better, equal, worse)
4. Interval ranking
5. For extrarational standards compare and ask if values: enforce, are neutral, or conflict?

Multidimensional scales (mixes of above)

Table 7.4 *Evaluation Exercise**

Basic Evaluation	Variant 1	Variant 2	Variant 3
(Read Top to Bottom for Each Variant)			
1. Identify the system			
2. The decisionmaker is: _____ _____	_____	If you were the decision-maker	Select a new decisionmaker
3. The client is†: _____	If the client change to: _____	Who would you judge the client to be?	Select a new client _____
4. The system's evaluation standards are‡: _____ _____ _____	What would the standards be? _____ _____	What would the standards be? _____ _____	What would the standards be? _____ _____

*This exercise is a variant of one described by C. West Churchman in "A Critique of the Systems Approach to Social Organizations," *Systems Concepts: Lectures on Contemporary Approaches to Systems,* Ralph F. Miles, ed. (New York: John Wiley, 1973), p. 201.

†The client can vary through a spectrum of: top management, middle management, supervisors, employees, stockholders, government, product users, public served or other as you designate.

‡Refer to the Systems Evaluation Table 7-3 for a list of standards and to Chapter 7.

from an optimal quality standard comparison; better than engines of its type made in the past; as good as the similar engine made by a competitor; not quite meeting the standards of our planned quality, but of satisfactory quality. Survival quality could be an applicable standard if our engine were destined for a military jet fighter that would be dueling with an enemy's jet. The quality demands of the clients would unquestionably be a standard as would government standards for safety and emissions. There could be legal quality standards involved in the production contract and certainly professional standards of the company, whose own survival was dependent upon delivering operational engines, would apply.

Without stretching the imagination too far some of the extrarational standards would also apply. There may have been political limitations put on funding, or internal political constraints involved in company organization, labor-management problems, or associated with the process of bidding to receive the contract. Energy conservation groups might consider jet engine production to fail to meet the ethical and moral standards of increasing consumption of the world's limited and irreplaceable fossil fuel supply. If the engine were produced for a new civil airline planning to transport thousands of tourists a year into previously remote south Pacific Islands, cultural standards could be a consideration for evaluation. Most readers might think that I have an example where creative and artistic standards would certainly not apply; however, my pilot colleague of many years, Bill Mol, used to speak of "painting the sky" when his fighter had the capability for ease of maneuvering that an engine with a high thrust-to-weight ratio provided, and for many fighter-pilots that metaphor recalls visual and kinetic experiences that fall into the artistic and creative realms.

An example of the multidimensional scales might be the statistical sampling of freeway traffic volume, the ratio scales of traffic accidents to freeway locations, and the political standard of "minimum fatalities" to form vectors for decisions in traffic control, safety, and maintenance. See Chapter 11 for an illustrative case study.

Saying that there are many standards that could be applicable for the evaluation problem at hand is different from saying they all should be. Selecting standards, as well as selecting secondary criteria, are meta-policy functions falling into policy sciences. Since human systems are too complex to evaluate all their components using all possible standards, judgment is indispensable. The importance of a checklist such as provided in Table 7.3 is that evaluation after explicitly rejecting some standards and selecting others is far superior to evaluation where standards are either implicitly rejected or accepted and where leadership is unable to fully explicate the rationale for its decisions.

EVALUATION AND MEASUREMENT SUMMARIZED

The fundamental statement concerning the Systems Analysis evaluation method is

---

IF SYSTEM QUALITY CAN BE DETERMINED USING VALID
CRITERIA, THEN IT CAN BE JUDGED IN COMPARISON WITH
STANDARDS FOR DECISIONMAKING

For a review of the evaluation process reconsider the corporation example. In Table 7.5 assume that the analyst-decisionmaker team is concerned about the net output of organizational effectiveness and health as the primary criterion and has concluded that it can best be measured through the secondary criteria of (1) productivity, (2) employee satisfaction, (3) environmental sensitivity, and (4) leadership. Furthermore it was decided that measurements would actually be made of cost per unit, units per time, and total sales for productivity; employee grievances and absenteeism for employee satisfaction; plant emission levels for environmental sensitivity; and participative management procedures and practices plus middle management attitudes toward participative management for leadership. The measurement techniques selected were observation, reports, and financial statements analysis, interviews, and survey questionnaires to answer the Step 1 question of "What quality exists?" Note that in this case there are tertiary criteria used through the assumption that measurement of cost per unit and unit per

Table 7.5 *Evaluation of Corporation "X"*

Primary Criterion: Organizational effectiveness and health (net output)

Secondary Criteria	Evaluation Step 1: Determine Quality	
	Method	Finding*
Productivity		
Cost/Unit	Analyze reports	$1/unit (1978)
Units/Time	Measure	100/day (1978)
Total Sales	Observe	Down 10%
Employee satisfaction	Study records	
Grievances		22 (1978)
Absenteeism	Interview	
	Observe	
	Survey	Low (1978)
Environmental sensitivity	Measure	Meets govern-
Emission levels		ment standards
	Observe	
Leadership	Survey	Top management
Participative decisionmaking	Interview	open and encour-
Management values	Observe	ages innovation
		Some middle-
		management values
		conflict

*All hypothetical.

time are adequate to determine the quality of productivity, which, in turn, is assumed to be one of the four major variables in organizational effectiveness. When tertiary criteria are used analysts and decisionmakers must be cognizant of the increased uncertainty involved and be especially careful to review the "good reasons" for assuming positive correlation with organizational effectiveness.

Let's further assume that Step 2 our judgment has produced the standards of past quality, quality of other similar systems, government quality standards, quality demands of the client and managerial values as being adequate to provide the answers to the question: "How good is that quality?" Given a larger research and

Evaluation Step 2: Judge Quality		Conclusions and Actions Required
Method (Standards)	Finding*	
Past quality	$1.25/unit (1977), "better"	Productivity is improved over last year, but sales are down
Client demands	80/day (1977), problem!!!	with inventory building Need marketing and R&D, check competition
Past quality	28 (1977)	Grievances are down compared with last year Absenteeism for all reasons remains low No apparent action required
Other system	Low (1977)	No data for other systems
Government standards	Satisfactory quality	Meet government standards but below the demands of
Client demands	Does not meet desired quality	environmental groups Study ways of improving
Creativity standards	Top management feels new ideas not being used	Overall leadership quality is satisfactory, but top management's desire for
Managerial values	Marginal quality Problem!!!	more participative management not being implemented Sales drop may be related Item for next management meeting

analysis capability our preference might be to explicitly investigate other standards, such as optimal quality and survival quality, but in this case management judges that it has a good intuitive feel that neither is necessary at this time. Business indicators are not so alarming that survival quality is a threat nor do they indicate such successful operations that a basic strategy decision involving raising goals to optimal quality is warranted now. If you are surprised at the degree of judgment required in this hypothetical example my comment is that this presents realistic policymaking requirements concerning the evaluation process. The analyst's expertise should be there to help management reduce the probabilities of elimination of a potentially critical variable before the analysis begins but there is no way of removing all uncertainty toward the adequate selection of criteria and standards.

The results of this evaluation, summarized as shown in Table 7.5, could form the basis of a systems analysis briefing to provide top management results of the study. If our hypothetical corporation were in the electronics business, the 10% sales drop, in spite of other positive indicators, might lead management into more research of other systems' products and increased R & D efforts toward new products. Top management would be faced with a problem of education, persuasion, or replacement of the portion of middle management who do not share their values toward participative decisionmaking by supervisors and employees. On the other hand, emergence of that values conflict might result in top management's reconsideration of participative management as a valid secondary criterion to measure organizational effectiveness.

Application of the systems analysis evaluation methodology helps to monitor indicators and trends, forces explicit consideration of alternative criteria and standards, encourages better measurement techniques, can overcome professional prejudices, biases, and narrow vision of the policymaking system, and becomes a tool to change political feasibility of previously rejected alternatives.*

8

Quantitative Tools for Systems Analysis

After stating views on the false dichotomy of any qualitative versus quantitative debate and summarizing the qualitative methods of systems analysis covered in Chapters 1 through 7, Chapter 8 provides the justification, knowledge requirements, fundamental quantitative management information model, specific techniques, and pitfalls involved in the quantitative dimension of systems analysis. The case studies in Chapters 9, 11, 16, and 18 illustrate the application of quantitative techniques and the axiom of "appropriate methods for unique problems."*

Conclusions drawn in the absence of quantitative measurement will have doubtful validity. On the other hand, action taken solely on the basis of quantifiable variables can lead to inappropriate, wasteful, noneffective, or undesirable results. The only utility of an "either-or" type debate over qualitative versus quantitative methods is for narrow pursuits of intellectual interest or rationalization of a cherished academic orientation tenaciously continued in spite of, or in the absence of, systems evaluation. For the solution of complex problems in human systems the debate has no relevance as the zero-sum model does not apply. Analysis must involve both qualitative and quantitative tools. Decisions always include both categories of variables—even if one or the other is poorly explicated.

Chapters 1 through 7 have provided an array of qualitative methodologies and techniques, some of which stretch the traditional parameters of systems analysis. They are summarized in Table 8.1. The systems analysis concept presented in this book departs from much previous systems analysis literature in that I do not maintain that a mathematical systems model must be the central component of the problem solution. The central component of analysis may be quantitative, qualitative, or mixed, which is the usual case. The idea that a mathematical model must be created to represent the system under study before the approach can be classified as systems analysis belongs to what could be called the "normal science paradigm" of systems analysis—referring to the Thomas Kuhn explanation of progress in science.[5] This difference, and others, places *Systems Analysis and Policy Sciences* in a new systems analysis scientific paradigm.

---

*I am indebted to Dr. George Jones, University of Southern California, for his critical analysis and improvement suggestions relating to this chapter in its earlier draft form.

Table 8.1   *Qualitative Tools in Systems Analysis: A Review*

Macro Tools:

The "policy sciences—systems analysis" paradigm with foundations in scientific knowledge, the scientific method, the systems approach, and management sciences (Chapters 1, 3, and 5)

Theory and models as catalysts for systems knowledge (Chapter 2)

Levels and pitfalls of systems analysis (Chapter 4)

Policy strategies, policy analysis, policymaking models (Chapter 5)

Systems analysis research methodology (Chapter 6)

Tools

Feasibility analysis (Chapter 6)

Systems analysis evaluation and measurement model (Chapter 7)

Values analysis (Chapter 5)

Extrarational variables analysis (Chapter 5)

Political variables analysis (Chapter 5)

Cross-cultural variables analysis (Chapter 5)

The systems analysis briefing (Chapter 6)

Future studies (Chapter 5)

The objective of including explicit qualitative methodologies is to provide analysts and decisionmakers with rational means for the inclusion of those qualitative—and often extrarational—variables in the analysis. The problem for the analyst is not whether those qualitative variables exist in human systems. They do exist. The problem is how to more rationally consider those variables on the assumption that the alternative is for them to remain implicit and unanalyzed while they continue to move systems toward quality improvement or deterioration. If their influence is toward improvement, we would like to know more about the functioning of those variables so the process can be continued, expanded, or duplicated in other systems. If the influence is toward deterioration we want a capability for identification, evaluation, and rectification before system failure becomes irreversible. We cannot cure an unknown system pathology. One lesson of the 20th century is that hidden pathologies tend toward system breakdown rather than self-amelioration.*

The quantitative-versus-qualitative analytical techniques argument is truly a false dichotomy—both are necessary and neither alone is sufficient. You will note that every case study in Part II relies on both qualitative and quantitative argu-

---

*See Part III, Chapter 19 for a discussion of the implications for decisionmakers and scientists of the implied assumption here that the system under study should be improved rather than allowed to fail.

ments for its recommendations and findings. Often the mix of qualitative and quantitative are interdependent in an evolutionary manner. For instance, a values orientation toward alleviating hunger for peoples throughout the world has resulted in allocation of resources to measure the dimensions of the problem and to accomplish research into possible system improvements. The results of those measurements both reinforce the tacit knowledge (i.e., extrarational and non-scientific) that there was, or would be, a serious problem or an improvement possible and provided the quantitative base for specific rational actions to improve food production and distribution systems. Spencer's research in Chapter 13 shows that tacit knowledge existed in the 1930s, well before world hunger problems were an issue. His initial quantitative and scientific research, which proved the technological feasibility of soil-penetrating phosphates was accomplished by 1946. Political feasibility constraints to 1980 have prevented the continued research necessary to demonstrate the economic feasibility of producing a fertilizer that will penetrate below the plow sole. The Spencer case is a good example of two systems analysis principles. First is the interdependence of qualitative analysis and tacit knowledge in the creative process with quantitative research—in this case into improved plant production potentials; and secondly, the principle of the necessity for all three feasibilities—economic, technological, and political—to exist before impact on the policymaking system can occur.

The principles apply universally. To argue that the current improved understanding of the world hunger problem over that of 20 years ago, or 50 years ago, has resulted from quantitative analyses alone is a self-delusion. To attempt to identify the respective roles of, and contributions of, quantitative analyses and of qualitative analyses is a legitimate scientific area of research, itself, for systems analysis.

WHY QUANTIFY?

In order to accomplish behavioral research into "what is?" (see Chapter 6) and answer the questions of What? When? How Much? and How Many? we must measure in both qualitative and quantitative dimensions. Quantitative measurement allows us to organize knowledge, to be specific, to compare amounts over time and with other specific amounts which is the key to the two-step evaluation model of Chapter 7. It allows aggregation and manipulation of data in management information systems using ever more powerful, rapid, and accessible computers.

Quantification allows reduction of complexity and uncertainty to understandable levels for decisionmaking. It provides justification for stipulated system output. Through quantification we can record events for later review, evaluation, comparison, and validation within the scientific method. We can design feedback mechanism for control and decisionmaking by quantifying reports, relationships and events over time. Simulation is possible with quantification. This requires a structured, rational, and repeatable process capable of sensitivity analysis by adjusting quantitative independent variables for analysis of alternative outcome arrays.[38] Useful relationships can be observed through mathematical and statistically derived transfer functions. We cannot avoid addressing the issues of value (utility) in human systems along with the issues related to values questions. Quantification is necessary, although not sufficient, to measure system efficiency, effectiveness, and quality. Systems analysis is impossible without some degree of quantification. The ability to abstract the physical and social world into quantifiable models has been a major contributor to scientific, technological, and social progress. In summary, we must quantify to understand or improve our human systems.

KNOWLEDGE REQUIREMENTS
FOR QUANTIFICATION

Quantitative techniques in systems analysis apply logic and reasoning with the help of economics, mathematics, computer sciences, statistics, probability theory, and decision theory to assist in decisionmaking. The mathematics involved includes linear and nonlinear mathematics, Boolean and matrix algebra, set theory, network theory, and portions of geometry and calculus. Computer sciences knowledge involves various types of programming and use of algorithms plus the design and utilization of management information systems (MIS). The primary criterion is for the application of quantitative technique to increase the decisionmakers ability to make the correct decision. Reasoning, using quantitative models, is deductive.[39] Experience is inductive and the judgment necessary to select the right quantitative techniques or combination of techniques for the unique system problem results from a combination of tacit and explicit knowledge possessed by the system analyst(s) and the decisionmaker(s).

FUNDAMENTAL QUANTITATIVE DECISION MODEL

Table 8.2 graphically portrays the fundamental quantitative decision model. The components of the model are the decision alternatives that could be made (D_1 through D_n); the states of the world which could occur (S_1 through S_m); the likelihood, or probability, of these states actually occurring (ℓ_1 through ℓ_m); the outcomes (O_{11} through O_{nm}) which could result from the application of each decision or alternative in each state of the world; and, finally, the payoff or worth (W_{11} through W_{nm}) or value or utility (all synonymous terms) associated with each outcome. Outcomes and worths may be positive (where the decision strategy is to maximize gain) or negative (where the decision strategy is to minimize loss). If there is only one state of the world in our analysis we are dealing with certainty or deterministic models. If there is more than one relevant state of the world that can occur, then each decision can result in as many outcomes as there are states of the world and we are dealing with uncertainty or probabilistic models of decisionmaking. Management science and systems analysis have developed specific techniques for dealing with both kinds of problems.

Table 8.2 *Fundamental Quantitative Decision Model*

States of the World:	S_1	S_2	...	S_j	...	S_m
Likelihood of States of the World:	ℓ_1	ℓ_2	...	ℓ_j	...	ℓ_m
Decision Alternatives						
D_1	O_{11}/W_{11}	O_{12}/W_{12}	...	O_{1j}/W_{1j}	...	O_{1m}/W_{1m}
D_2	O_{21}/W_{21}	O_{22}/W_{22}	...	O_{2j}/W_{2j}	...	O_{2m}/W_{2m}
$\vdots$	$\vdots$	$\vdots$	$\vdots$	$\vdots$	$\vdots$	$\vdots$
D_i	O_{i1}/W_{i1}	O_{i2}/W_{i2}	...	O_{ij}/W_{ij}	...	O_{im}/W_{im}
$\vdots$	$\vdots$	$\vdots$	$\vdots$	$\vdots$	$\vdots$	$\vdots$
D_n	O_{n1}/W_{n1}	O_{n2}/W_{n2}	...	O_{nj}/W_{nj}	...	O_{nm}/W_{nm}

As an illustrative example under certainty in a deterministic inventory type problem let's assume the decision (D) to be made is the ordering frequency and lot size of microbus overhauled engines for an imaginary Krone Trans-Asian Touring Company and our interest is in minimizing the total relevant costs, in dollars (as W), when the rate of needed overhauled engines is known from past experience and remains constant. Our state of the world (S) under conditions of certainty, then, is the state where the rate of required overhauled engines remains constant throughout the year and the total number needed is derived by multiplying kilometers driven per year per bus (50,000 km) by the number of buses (100) and divided by the average kilometers driven per required engine overhaul (34,722 km), which provides the overhauled engines per year required (units required or demand) equaling 144 overhauled engines (business is good!).

QUANTIFICATION OF THE PROBLEM

Let C_o = ordering cost given (1) \$10/engine in the order, and (2) plus \$150 per order for handling/receiving. Also let C_n = cost of holding a unit of inventory per unit time given an average of \$60 per engine per year for storage, insurance, security, and interest on engine cost. Finally let D_n = annual demand given a known 144 engines per year. S_1 = state of the world.

We are solving for: (1) Q = lot size per order (optimum); and (2) D_n/Q = frequency of orders/year. Then

$$Q = \sqrt{\frac{2(150)D}{C_n}} = \sqrt{\frac{2(150)(144)}{60}} = 26.83 \text{ Lot size or 27 engines} \qquad (1)$$

And

$$\frac{Q}{2} = \text{average inventory (uniform usage)} = 13.415 \text{ or 14 engines} \qquad (2)$$

And

$$\frac{D_n}{Q} = \frac{144}{27} = 5.33 \text{ optimal ordering frequency per year} \qquad (3)$$

Note: To simplify the operation we could order five times per year to meet our annual demand, or every ten weeks, by adjusting our lot size for the first four orders to 29 engines and the size of the fifth order to 28 engines and accept a somewhat less than optimum solution to meet our demand.

With that lot size and frequency we can then compute optimum total cost for holding and ordering of engines (exclusive of the cost of the engines, which we have assumed as a constant) through the formula:

$$\text{total relevant annual cost} = C(Q) = \frac{Q}{2}(C_n) + \frac{D}{Q}(150) + \frac{D}{Q}(10 \times Q)$$

$$= \$3,050.00 \qquad (4)$$

Other calculations can show that this solution produces minimum annual cost with our problem variables.

We can now place in our decision model the total relevant inventory cost for the year of Trans-Asian Touring Company overhauled engines at $3,050, which becomes the "Worth (W_{11})" after the "Decision Alternative (D_1)" under the "State of the World (S_1)." Remember we have only one state of the world given in this deterministic decisionmaking model (i.e., "144 engines per year demand at uniform usage with no safety stock"). Futhermore we assumed a likelihood of that state of the world occurring as being 100%, of certainty. A second decision alternative (D_2) might well be that we are uncomfortable with our assumption in the decision alternative, D_1, that our inventory should be completely exhausted at the end of every 10 week period (i.e., a lead time of zero) and that we want to consider a more conservative case where we maintain two overhauled engines in the inventory over and above the average requirements as a hedge against engine breakdowns that might occur at random times. For decision alternative D_2, then, we must consider the holding cost of two extra engines. That total cost would be

$120 ($60 per engine per year × 2 extra engines). Our capital investment will also go up by the cost of the two engines. Our decision now is between a high-risk alternative (D_1) with no inventory reserve and a low-risk one requiring additional holding costs and capital investment.

The Fundamental Quantitative Decision Model in Viewgraph 8-2 also allows us to move our analysis from a deterministic model (the one we just examined) to a probabilistic model where there is more than one state of the world. We could have assumed that random engine overhauls would be required at a rate per year of 150 engines as a second state of the world (S_2). We could also consider the possibility that our business would expand during the last half of the year to a point where two-hundred engine overhauls would be required and consider that state of the world as, S_3. For each of these states of the world we would need to compute our estimates of how likely those states were to occur in the future and include those in the blocks, ℓ_1, ℓ_2, and ℓ_3. Furthermore we could identify a high-risk and a low-risk decision alternative for each of those states of the world. Probabilistic quantitative decision models could then be used to calculate the outcomes and the worths associated with each decision alternative and each state of the world.

In either case—the deterministic or the probabalistic decision models—the principle is the same. Quantification provides the comparable data for alternative outcomes (Os) with their associated worths (Ws), which act as assists in the decisionmaking process.

I have carefully selected the word "assist" as it must be obvious from even this simple problem that the decision cannot be based on quantification alone. Obviously my Trans-Asian Touring Company must compare relative costs and consider as many hidden costs as possible such as costs of handling, insurance, storage, accounting, security, and deterioration over time and how those costs might vary in the future with inflation or rising labor costs. Quantification in systems analysis is essential for those purposes but the intangibles and extrarational variables are always there to be handled by qualitative analysis. Perhaps the President of the engine overhaul company I am using in Korea is my good friend Mr. Sung Ho Lee who has some excellent contacts within Korea and this association is helping greatly in the expansion of my touring business in that country as well as giving me sound advice on business diversification possibilities into charter air flights on a joint venture basis. Furthermore, my wife, Sue, loves to shop in Korea. A suggestion had been made to move our overhaul business close to the home office in Singapore or perhaps to the Philippines where labor costs are less than in Korea. But that would remove the possibility of frequent trips to Korea to shop and visit our friends, the Lees. Such extrarational considerations must be weighed as worth (W) along with the worth data obtained from the quantitative systems analysis. Making money for the business is one thing, the quality of life is another, and both become real variables in the decision process.

We can place a quantitative weighting on extrarational variables for purposes of calculation within the decision model. For instance in computing the worth (W) of a decision alternative (D) we can identify the variable of "shopping and friends in Korea" and the variable of "possible future gains through Korean business contacts" and decide that each should carry twice the weight (measured in dollars) of our purely economic variables such as engine ordering and holding costs, but that when the economic projections exceed double the extrarational variable figure (given a weighting factor $D \times 2$) we will agree to sacrifice our concepts of quality of life for the economic gain. The important factor to remember is what weighting you have placed on those extrarational variables in relation to your own wants and needs. The mathematical results will be heavily biased by those weightings that flow from values (see Chapter 5) and those that flow from value (utility) calculations.

I wish to make one more point with regard to the fundamental quantitative decision model before going on to list some specific tools and techniques for quantification in systems analysis.

If through our quantitative analysis, comparable figures for worth (W) can be developed for each outcome associated with each alternative decision within each possible state of the world, this will provide the quantitative answer for the question of Step 1 of the evaluation system outlined in Chapter 7 of "What is the Quality?" In this case, since we are dealing with forecast future outcomes rather than actual observed ones the precise question will be "What will the quality be?" The quantitative answer to the question of Step 2 of the evaluation system (i.e., "How good is that quality?" or, in this case, "How good will that quality be?") can now be obtained by the worths (Ws) determined with the standard of evaluation selected. You then have a normative model for decisionmaking. That is, you have computed all relevant outcomes (Os) and determined the probable quality resulting from those alternative decisions by also computing worths (W) and now selection of the right decision can be made using your selected standard. The standard of "optimum profit for the shareholders" will obviously dictate a decision from the array of outcomes and worths computed that will be different from the decision derived from using the standard of "quality demands of the clients" or some combination of extrarational standards. Selection of standards is another place where quantitative analysis and values analysis merge.

Quantitative tools of systems analysis provide the means to compute, within given assumptions, both the outcomes (O) and the worths (W) for the fundamental quantitative decision model. Since complex human systems always have inherent uncertainties the relevance of quantitative systems analysis is sensitive to the states of the world assumed as well as to the standards selected by the decisionmaker and analyst for judging the quality obtained (or forecast to be obtained in the future). I will refer back to the fundamental quantitative decision model later after discussing some specific quantitative tools.

SPECIFIC TOOLS AND TECHNIQUES FOR QUANITIFICATION IN SYSTEMS ANALYSIS

According to a standard systems analysis text of the 1960s "A principal aim of systems analysis is to find the relationship between cost and effectiveness."[40] Providing decisionmakers alternative courses of action based on the costs and benefits of each, quantitatively expressed, remains a basic concept of systems analysis. The fundamental quantitative decision model of Table 8.2 is the expected value type model.

The specific tools and techniques to accomplish quantification in systems analysis can be usefully divided into deterministic analytical models and probabilistic analytical models. As is the case in all such arbitrary distinctions made for educational purposes that distinction in real world problems is not a mutually exclusive one.[41] Furthermore in dynamic human systems there are very few exclusively deterministic type problems. Problems tend much more to resemble Winston Churchill's famous quote on the uncertainty of Russian intentions about entering World War II; "The answer . . . is a riddle, wrapped in a mystery, inside an enigma."

DETERMINISTIC ANALYTICAL MODELS

Deterministic models are those that are applicable to problems in which there is only one state of the world (*S*) and variables, constraints, and alternatives are,

Table 8.3 *Illustrative Deterministic Quantitative Models and Techniques in Systems Analysis**

Model, Tool, or Technique	Application	Knowledge Base
Linear programming	Allocation, distribution, and optimization in business, transportation, inventory, construction, logistics, and networks	Computer science, sensitivity analysis, algebraic solutions, simplex tableau, and economics
Queueing theory	Waiting/service ratios and people/things/events	Monte Carlo, simulation, and statistics
Program management techniques	Production and construction planning	PERT (cost or time), GANTT charts, network analysis (CPM), and decision trees
Markov analysis	Marketing/sales/forecasting	Matrix algebra and economics
Conflict analysis	Business, psychology, and security studies	Game theory
Quality assurance	Industry/defense	Technology and science
Cost/benefit	Resource allocation	Economics and statistics

*Deterministic Models are those which are applicable to problems where there is only one state of the world assumed and where variables, constraints and alternatives are, after acceptable assumptions, known, definable, finite, and predictable with statistical confidence.

after making acceptable assumptions,[42] known, definable, finite, and predictable with statistical confidence. Table 8.3 lists illustrative deterministic models and techniques to obtain quantitative knowledge in systems analysis.

UNCERTAINTY AND RISK

When our probability or likelihood (ℓ) assessed for a state of the world (S) to occur is less than one (1.0) we are working with something less than certainty— or uncertainty. Multiple possible states of the world, each with their own assessed occurrence probabilities add to our uncertainty. In these cases outcomes (O) are usually stated in probabilistic terms such as my scientific statements for behavioral and normative research categories in Chapter 6. When the likelihood of future states of the world can be specified then a decision under computed risk is possible. Risk is essentially the expected value of what loss could occur, but the expected value of positive outcomes are often included in this category as well.[43] A decision under true or complete uncertainty occurs when the likelihoods of states of the world cannot be specified—for whatever reason. Decision principles used in this case cover the spectrum of those policy strategies and alternatives to the systems approach that were covered in Chapters 3 and 5. Deterministic models of decision-making often assume away uncertainties or risk to allow for ease and conservation of analysis. Those assumptions may be warranted or not. The systems analyst has a responsibility to make the assumptions explicit so that a judgment of accepta-bility and desirability can be made about those assumptions. For instance, it may be perfectly warranted to use deterministic models to compute optimum inventory stock levels based on the assumptions utilized for last year's levels in the light of our behavioral research into what actually happened as a result of those assump-tions. As an unevaluated system in a changing environment will deteriorate, in the absence of that behavioral research and challenging of assumptions it may be completely unwarranted and dangerous. In cases of revolutionary, or even very rapid, system transformation all planning and systems analysis has doubtful valid-ity. However, setting that caveat aside, systems analysts have a broad spectrum of models and tools developed for dealing with uncertainty of the future. When uncertainty and risk are involved, probabilistic quantitative models of decision-making are employed.

PROBABILISTIC ANALYTICAL MODELS

Probabilistic models are those where there is more than one state of the world (S) and where each possible state must be estimated or defined [likelihood of occurrence (ℓ) in Table 8.2] to allow computation of the conditional outcomes (O) of each decision alternative (D) in each state. Alternatives may be numerous. Mathematics, statistical inference, and probability theory are used to reduce un-certainty within acceptable assumptions. Probabilistic analytical situations can be reduced for deterministic model applications by assuming the one most likely

Table 8.4 *Illustrative Probabalistic Quantitative Models and Techniques in Systems Analysis**

Model, Tool, or Technique	Application	Knowledge Base
Dynamic programming	Multistage decision in production, allocation	Computer science and probability theory
Computer simulation	Systems interactions	Computer science and Monte Carlo
Probabalistic inventory models	Where demand and/or lead time are random	Probability theory and expected value statistics
Stochastic models	Computing probabilities of system transition	Matrix algebra and calculus
Sampling, regression, and exponential smoothing	Problem solving with large populations	Statistics and probability theory
Bayes theorem	Forecasting under conditional probability and dependence, causal analysis	Algebra, probability theory, and knowledge of prior probabilities
Cost/benefit analysis	Resource allocation	Economics and statistics
Fault tree analysis	System behavior	Algebra and statistics

*Those quantitative models where there are more than one states of the world and where each possible state must be estimated or defined to allow computation of the conditional outcomes of each decision alternative in each state.

future or by analyzing only the worst state or the best state identified. Probabilistic models can also analyze the interactions of several states of the world under varying assumptions such as the dynamic models used in the Limits to Growth (later termed the Alternatives to Growth) investigations between the five major trends impacting on world stability; population expansion, industrial growth, consumption of nonrenewable resources, agricultural production, and pollution. Probabilistic models hold the most promise for system improvements through quantitative analysis. Table 8.4 lists illustrative probabilistic models, their applications and respective knowledge bases. Chapter 18 is a relevant case study.

COST ANALYSIS

As the reader will note from Tables 8.3 and 8.4 cost analysis is included in both deterministic and probabilistic types of quantitative decision models. It is the oldest and most common category of analytical models. Cost compared with measures of effectiveness was, and still is, the backbone of systems analysis in defense planning, industry, and business.[44] As system resources are invariably reduced to dollars or dollar equivalents, as budgets and financial reports are expressed in dollars, and as success or failure in a large percentage of human systems hinges on income meeting or exceeding expenditures, cost analysis—through economic models of one form or another—will continue to be a standard tool for the systems analyst.[45] The concept of utility as an alternative to money, and that utility is not linear to money, as well as many other cost analysis concepts such as opportunity cost (or loss), expected value, conditional profit, and cost sensitivity given varying scenarios are included in those cost analysis models. References containing detailed instructions on application of the models and techniques shown in Table 8.3 and Table 8.4—and other techniques—can be found in the Bibliographic Essay. Many of the quantitative techniques require a complete text to be adequately described.

Table 8.5 *Potential Pitfalls in the Use of Quantitative Techniques*

Adapting the problem and the real world to fit the formula (the Procrustean metaphor)

Model reification (fascination and preoccupation with details of the model while being overcome by events in the real world—"seeing the trees and not the forest" metaphor)

Ignoring the axiom of "appropriate methods for unique problems"

Overconfidence and oversell

Oversophisticated models and techniques requiring non cost/benefit allocation of resources

Using the wrong model for the problem

Using the right model incorrectly

Tautological solutions (those highly sensitive to the statement of the problem and methodology employed)

Making a butch (Herman Kahn's "completely mistaken technical notion or fact"*)

Interest in only the worth (economic utility) of the outcome and not the values of the system

De-emphasis or ignoring of qualitative or extrarational components due to reliance on mathematics and associated rational models of policymaking

Overuse of technical and mathematical language—thus failing to communicate

---

*See Chapter 5, reference Note 31 in Part IV.

PITFALLS AND LIMITATIONS OF QUANTIFICATION

Quantification of variables involved in complex problems has a both real and an imaginary capability to reduce the uncertainties of decisionmaking. The art of quantitative systems analysis lies in the analysts' and the decisionmakers' ability to judge which uncertainties are being removed advantageously and, conversely, which ones fit the Procrustean metaphor. There is no one formula for that judgment. Correct application of quantification will usually make the difference between success or failure in reaching system goals through data manipulation to (1) assist in selecting preferable policy alternatives or (2) in the entire separate category of problems where policy implementation to achieve those goals is the aim. Just as there are pitfalls in the use of systems concepts (Table 3.2) and the use of systems analysis (Table 4.3) there are also pitfalls when utilizing quantification in systems analysis. Table 8.5 lists the essential ones. The merging of computer sciences with management sciences and policy sciences has opened vast new realms of possibilities for the advancement of all three categories of knowledge—environmental, human, and control (see Chapter 1). With the exponential increase of system interdependence as science and technology continue to produce a relatively smaller planet earth, the social impacts of mistakes in our calculations or gaps between available knowledge and knowledge applied to improve human systems also rises exponentially. Those mistakes and knowledge gaps can be either qualitative or quantitative or a mixture of both. The professional systems analyst has a responsibility to be aware of the limitations and pitfalls associated with the current state of the art as well as with the potential benefits. Part II, following, provides practical application of several quantitative models and techniques.

Part II

PRACTICE

The test of systems analysis theory is in its application to problems and in its ability to improve human systems. Systems analysis often involves and requires teams of researchers working over long periods of time with highly sophisticated computer-based data sources. Such was the case, for instance, with a Japanese firm that made an extensive 2 year feasibility study before the decision to build a multibillion dollar ship-building and repair facility in Subic Bay, Republic of the Philippines commencing in 1979. Part II of this book demonstrates that large consulting firms, think tanks, or policy institutions are not necessary for systems analyses that meet the test of the first sentence of this paragraph.

Chapters 9 through 18 are in most cases, abbreviated versions of longer studies. The systems analyses in Chapters 9, 10, 14, 15, 16, and 17 were accomplished by graduate students in the University of Southern California Master of Science in Systems Management (MSSM) degree program as research papers for the 8-week SSM 665 Systems Analysis course. The authors are all midcareer professionals pursuing the MSSM degree during evening courses while carrying on their normal work responsibilities. The conclusions and recommendations are those of the individual authors. The methodologies utilized reflect various elements of the systems analysis theory presented in Chapters 1 through 8 and are summarized for the reader in the introduction to each case study.

9

Case Study in Pollution Reversal Systems:Removal of Abandoned Wrecked Vehicles

CLARENCE H. WALKER, JR., C.M.

Clarence H. Walker, Jr. wrote this study in the Spring of 1977 for a USC/
MSSM course at Kwajalein Island, U. S. Trust Territory of the Pacific. At
that time, he was Chief, Plans and Programs for Global Associates.
  The study illustrates:

1. A systems analysis of a whole system (see Table 4.2, Level 4).
2. A Pareto Optimal solution. That is, an action that will result in a net
   benefit for all parties concerned.

The system proposed by Mr. Walker has a patent pending and a prototype
vehicle is under construction.

R. M. KRONE

INTRODUCTION

Millions of vehicles are abandoned/wrecked each year in the United States. Some
of this abandonment/wreckage is temporary, and some of it is permanent. Normally, it is the result of a mechanical failure or an accident. These vehicles are seen
on the shoulders of highways, in ditches along the roadway, in fields, in peoples'
front yards and backyards, along crowded streets, and in any number of other
locations. The vehicles are in various states of deterioration, in grotesque shapes
as a result of accidents, and/or in states of disrepair due to individuals having
stripped them of parts. Some are just rusted hulks, some have grass and weeds
growing through them, and for others the reason for their abandonment is not
obvious. These vehicles destroy the aesthetic appearance of the landscape. Many
are extreme safety hazards having even been the cause of other accidents. However,
all of these vehicles appear to have one thing in common—the debasement of our
environment.

Why are they there? Probably in many cases, it can be attributed to the affluency of our society. However, there are a multitude of reasons. One attitude is that if it quits on the highway and it's not worth fixing, then leave it there. There is often no law or penalty against it. The individual usually has the means to obtain another vehicle or may do without one for awhile. It's not worth the inconvenience, nor the time, nor the cost to take it to a junk yard. If it is in its present condition because of an accident and has been "totalled," using a common vernacular term, then an insurance company has probably assigned it to a junk dealer. Whether the junk dealer ever actually removes the vehicle is not the worry of the owner.

Whether by design or by fate, millions of vehicles are abandoned and/or wrecked each year. As an example, according to Mr. Butler, the Traffic Commissioner of New York City in 1972, there are approximately 86,000 vehicles abandoned annually in metropolitan New York. In fact a rough order of magnitude (ROM) estimate is that 2,000,000 vehicles are abandoned each year in the United States and another 8,000,000 vehicles are wrecked to the degree that they cannot be driven away from the scene of the accident. This amounts to approximately 10,000,000 abandoned/wrecked vehicles in the United States annually. If that figure appears staggering, multiply it by the cost to remove each vehicle (approximately $100 each in 1979) and the dollar value alone reaches one billion dollars annually. Add to this the number that are abandoned/wrecked in all the other countries of the world and the figure staggers the imagination.

THE PROBLEM

Most vehicles are left because a problem exists in the transporation/removal of the vehicle from the roadway or wherever the last point of nonlocomotion occurred. To illustrate the problem the following quotes are from newspaper articles in Hawaii, Massachusetts, Kentucky, and Australia:

In a report to the City Council, Mayor Frank F. Fasi said that while 600 to 700 vehicles are being abandoned each month, less than 400 are being collected some months. Police say one of the reasons these cars have been sitting so long is that the crane used to lift them away has been broken down. An average of three weeks elapses from the time an abandoned vehicle is reported until it is towed away. It often takes longer if the vehicle lacks some wheels and the crane has to be used.

From the Honolulu Star-Bulletin, June 1971

Five men and two wreckers supplied by the 84th Engr Bn spent two full weeks in Honolulu dragging cars from fields, empty lots, streets, and other places where many had lain for years. The men picked up 250 cars from the area and the project aim is to collect over 3,000.

From the Army Times, September 1972, Schofield Barracks, Hawaii

Abandoned and junked cars are becoming scarce here, thanks to the efforts of concerned National Guardsmen. Under the direction of Lt. Col. Richard Jordan, about a dozen men of the 26th Supply and Transport Battalion of the Massachusetts National Guard Division, using four dump trucks, a 20-ton crane, and two jeeps have been fanning out through surrounding communities in search of the eyesores. One small community logged 96 abandoned cars within its borders.

From the Army Times, July 1971, Framingham, Massachusetts

Members of the 29th's 594th (medium Truck) Company, along with forklift drivers from the 561st Maint Bn, have already removed 3,266 junk cars from a six-county area in Kentucky and Tennessee. This is a Domestic Action project and it gives the men a sense of accomplishment when they clean up an area. The work is being accomplished by two seven-man teams with five-ton flatbed trucks and forklifts. The civic group which assists in spotting the cars is paid a nominal fee for each car by the county. Out of this fee, the civic group (Boy Scouts, Jaycees, 4-H) pays for the lunches of the men and still has a little left over for their organization. Everyone benefits in one way or another from the project.

From the Army Times, July 1973, Fort Cambell, Kentucky

An alderman claimed yesterday that illegal car dumping in streets was costing New South Wales rate-payers $1 million a year. The Ashfield Mayor (Alderman P. Whelan) strongly criticized the State Government for failing to help combat the problem. It is a growing financial burden on councils and a constant source of annoyance for residents. Alderman Whelan said that councils were having to spend increasing time and money on the disposal of derelict vehicles.

From the Australian, August 1972, Sydney, Australia

Some vehicles have been damaged so severely in accidents that they cannot be easily towed away by a tow-truck. In those cases numerous pieces of heavy equipment must be used. This necessitates the utilization of cranes, trucks, cherry-pickers, and/or truck tractors with extra low-bed trailers (low-boys). Such a system of equipment is extremely large, slow, cumbersome, and very expensive.

This study will be limited to the fifty States of the United States and assume that six million of the total ten million abandoned/wrecked vehicles per year must be removed by some means other than a tow-truck system. In other words, the vehicles cannot be driven or towed away.

PRESENT REMOVAL SYSTEM

Currently, a system* (let's call it System A) of cranes and truck tractors with low-boys is normally used to remove the vehicles in the category established above.

*There are numerous combinations and types of equipment that are used, but this is probably the most common system.

The disadvantages of System A are numerous. System A equipment requires a large area in which to operate, it is slow moving, and it is awkward and cumbersome. In some cases special transporation permits are required to authorize it to operate on the streets. System A is very costly in both equipment and manpower (Table 9.1).

It requires four men to operate each unit of System "A". Using an average hourly wage of $12 per hour (1977 data) for each two operators and $8 per hour for each assistant operator (two required), the labor cost to operate one unit of System "A" is $320 per day. However the greater disadvantage of this type of removal system is the time element. Three examples of the time involved are listed below (source: newspaper accounts):

Example 1: 80 hours to move 20 vehicles = 4 hours/vehicle
Example 2: 560 hours to move 250 vehicles = 2.24 hours/vehicle
Example 3: Two 7-man teams, one year, to move 3266 vehicles = 3 hours/vehicle.

Based on information available and experience with heavy equipment it takes approximately three hours to move a derelict vehicle from Point A to Point B.

For this study, an optimistic time-study model of the operation of a unit of System "A" will be used to demonstrate the time element (Table 9.2).

At the rate of 100 minutes per unit operation a 4-man crew moves 5 vehicles per day (480 minutes/day) at a labor cost of $64 per vehicle (the 4-man crew cost is $320 per day). By applying this cost figure to the 6,000,000 vehicles of the study abandoned/wrecked every year the labor cost equals $384,000,000 per year. Now, let's add the equipment cost. Assuming that there are 260 work days per year (52 weeks times 5 days/week), then one unit of System "A" would remove 1,300 vehicles per year (5 per day times 260 days/year). Therefore, from a theoretical standpoint, it would require 4615 man-machine units of System "A" per year to remove the 6,000,000 vehicles. The capital investment of each unit of equipment (crane and truck-tractor with low-boy) is $98,000. Multiply this figure by 4615 units and it equals slightly over $450,000,000. However, by amortizing the total

Table 9.1 *Cost of Heavy Equipment with Manpower Requirements*

Equipment	Cost (each)*	Operators	Assistant Operators
Crane, 20-ton	$66,000	1	1
Truck Tractor	$24,000	1	1
Low-boy Trailer	$ 8,000	–	–
Totals	$98,000 per unit[†]		

*SB 700–20, Department of the Army Supply Bulletin, Washington, D.C.
[†]One unit consists of one crane and one truck tractor with low-boy.

Table 9.2 *Time Study Per Unit Operation (System "A")*

$$\text{Point A} \ \frac{(30 \text{ minutes})}{(10 \text{ miles})} \ \text{Point B} \ \frac{(30 \text{ minutes})}{(10 \text{ miles})} \ \text{Point A} = 60 \text{ minutes}$$

Time to load abandoned/wrecked vehicle on to the carrier = 30 minutes

Time to off-load abandoned/wrecked vehicle from the carrier = 10 minutes

Total: = 100 minutes

cost of equipment over a three-year period (the normal amortization period of this type of equipment), the annual cost of equipment would be approximately $150,000,000 per year. In addition to the labor and equipment cost, it is necessary to add the cost of operating these 4615 units of equipment. This covers fuel and maintenance including repair parts and a rough cost estimate is $70,000,000 annually. Therefore, the total costs for equipment, labor, and operation would be approximately $604,000,000 per year to remove 6,000,000 vehicles from various locations of our environment by utilizing System "A". This equates to approximately $101 per vehicle in 1979.

SYSTEM IMPROVEMENT

A system is required that will increase the efficiency of removing abandoned/ wrecked vehicles from the environment. The objectives of this new system are that it must be simpler, faster, relatively smaller, and more economical. I have invented a new system to meet all of these objectives. For this study it is called a Special Load-Carrying Vehicle or a "Forklift Truck." Essentially, I have married a truck and a forklift, by making very specific modifications to each. The invention* (pictured in Figure 9.1) is a single unit, load-carrying vehicle, having a motor-driven wheel-mounted chassis including a pair of longitudinal members which at one end (normally the front end) are rigidly interconnected, but from the other end (and for a considerable distance) are without transverse interconnection. There is a lifting fork provided between these longitudinal members with means for raising and lowering it so that the vehicle may manuever with the lifting fork lowered and sustaining a load in the lowered position. The fork may then be raised so that the load will be carried between the longitudinal members of the chassis. These are sturdy parallel longitudinal members, spaced sufficiently to be able to receive a vehicle between them and are laterally interconnected at the front only, where the engine and driver's cabin of the vehicle are mounted.

The vehicle is mounted on conventional wheels—the front wheels being driven

*Application for Patent, Provisional Specification, "An Improved Load Carrying Vehicle, "Patent Application No. PA–9767."

Figure 9.1 Special load carrying vehicle (forklift truck).

and steerable while the rear wheels are carried on short stub axles at the rear of the longitudinal chassis members.

A transverse fork mounting frame extends vertically up from the chassis close behind the driver's cabin of the vehicle, and a pair of parallel slide guide frames are secured rigidly to, and extend vertically up from, the two longitudinal chassis members, between the transverse fork mounting frame and the rear of the vehicle. A fork carrier is mounted in a vertically slidable manner on the fork-mounting frame. Then there extends rearward, from the lower part of the fork carrier, a pair of lifting fork arms. Means are provided for raising and lowering the lifting fork for example, an arrangement of hydraulic ram or rams, chains, and pulleys.

Preferably, the lifting fork and its actuating mechanism are such that the fork arms are tilted when the fork is raised to incline upward toward the rear. This tilting may be effected by one or more hydraulic rams connected between the fork arms and the carrier. The fork arms will be pivoted to the carrier about a transverse axis with stops limiting the pivotal movement of the fork in both directions. Alternatively, the mounting of the carrier for vertical movement on the fork-mounting frame may be such that when the carrier is fully lowered, it tilts somewhat to bring the fork arms horizontal and close to the ground. When the carrier is raised, the fork arms are at the same time brought to an upward tilted position. The carrier may travel in curved guides on the fork-mounting frame so that when it is raised from a fully lowered position, the fork arms are progressively inclined from horizontal to a tilted position.

Means may be provided for supporting, or locking, the lifting fork in raised position. Such means may be stops that are swung from the fork arms to engage with the side guard frames, or swung from the side guard frames to engage the fork arms, or some automatically engaging means, such as spring-loaded plungers.

In using the vehicle to pick up an abandoned/wrecked car, the lifting fork is lowered to full extent, and the vehicle is backed to cause the fork arms to run under the car. If desired, the fork arms may be fitted with rollers to facilitate this movement. The car is then located between the longitudinal members of the vehicle chassis. The fork is raised and tilted to lift the car well clear of the ground and retain it in place. Then the fork is locked in raised position, and the forklift truck may be driven away.

Load-carrying vehicles, according to the invention, will be found to be very effective in achieving the objectives for which they have been devised. The particular components of the invention may be subject to modifications of constructional detail and design, however this description suffices for the overall concept of the Forklift Truck.

THE SYSTEM B REMOVAL SYSTEM

The new system, called System B, has many advantages over System A. It is very simple to operate, it is faster, it consists of only one piece of equipment, and it is much more economical. It takes only a two-man crew. The assistant operator is primarily for safety purposes as the vehicle can be completely operated by one person. However, for this study, a two-man crew assumption will be made. By virtue of a two-man crew, this immediately indicates a reduction in labor cost for the operation of System "B" by 50 percent as opposed to the labor cost of System "A". However, there are additional considerations prior to arriving at the final cost of labor for this system. The most important is the time element for the removal of abandoned/wrecked vehicles by utilization of System B. The one man-machine unit of System B removes nine vehicles per day (Table 9.3). Therefore, the number of System B units required to move the 6,000,000 vehicles is only 2,564. This is 2,051 fewer man-machine units than required for System A.

At the above rate, a two-man crew moves nine vehicles per day (480 minutes/day) at a labor cost of $18 per vehicle (the two-man crew cost is $160 per day). By applying this cost figure to the 6,000,000 vehicles abandoned/wrecked every year in the United States, the total labor cost equals $108,000,000 annually, a tremendous reduction in labor costs when compared to System A (Figure 9.8). Now, let's add the equipment cost. Again assuming that there are 260 work days per year, then one unit of System B would remove 2340 vehicles per year (260 days times nine/day). It would require only 2564 man-machine units of System B equipment (the Forklift Truck) is estimated at $20,000. Multiply this figure by 2564 units and it equals $51,280,000. By amortizing this over a 3-year period,

Table 9.3 *Time Study Per Unit Operation (System B).*

Point A $\dfrac{(15 \text{ minutes})}{(10 \text{ miles})}$ Point B	$\dfrac{(20 \text{ minutes})}{(10 \text{ miles})}$	Point A	= 35 minutes*
Time to load abandoned/wrecked car on to Forklift Truck			= 10 minutes
Time to off-load abandoned/wrecked car from Forklift Truck			= 5 minutes

*An average estimated time dependent upon community size and traffic density.

the annual cost of equipment would be slightly in excess of $17,000,000 per year. In addition to the labor and equipment cost, it is necessary to add the cost of operating these 2564 units of equipment. This would cover fuel, oil, and maintenance to include repair parts. A rough order of magnitude estimate would be $20,000,000 annually. Therefore, the total cost for equipment, labor, and operation is approximately $145,000,000 per year to remove 6,000,000 vehicles from various locations within our environment by utilizing System B. This equates to approximately $25.00 per vehicle.

COMPARISON OF THE SYSTEMS

The comparison of the two systems reflects dramatically the advantages of System B over System A (see Figures 9.2, 9.3, and 9.4, and Table 9.4).

Through the use of System B the economic cost of removing the 6,000,000 vehicles per year from our streets and highways is reduced by 416 percent, or a savings of $459,000,000 annually—almost one-half billion dollars.

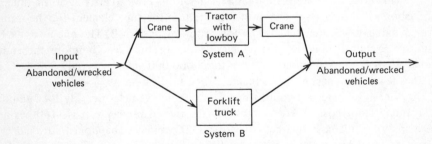

Figure 9.2 Input-output models.

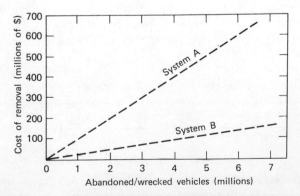

Figure 9.3 Cost—vehicle removal.

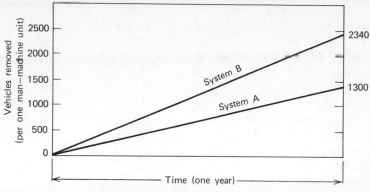

Figure 9.4   Time versus vehicles removed.

Table 9.4   *Comparison of Systems Costs (per year)*

		System "A"	System "B"
Labor		$384,000,000	$108,000,000
Equipment		150,000,000	17,000,000
Operating		70,000,000	20,000,000
	Totals:	$604,000,000	$145,000,000

DEVELOPMENT OF A NEW FORKLIFT TRUCK

Additional design work is required to develop exact specifications for the Forklift Truck. Next a prototype vehicle must be tested to ascertain whether it meets all of the specifications and objectives for which it was designed. Based on those results fabrication and marketing decisions can be made. For the fabrication and production, there exist at least two avenues of approach: (1) turn the specifications over to a company that currently has the capability and capacity to produce such an item in the quantities required, or (2) build an entire new plant and produce the end product by acquiring the resources (components) via subcontracts/service agreements. It is estimated that the development from design to completion of evaluation testing will take one year and cost approximately two million dollars, a small investment considering the potential of the system. All of the methodology of design and mechanical operation is within the present day state of the art. Based on the assumption that one forklift truck would be required for each residential area of 7500 individuals, then it is estimated that the initial potential U.S. requirement is approximately 12,000 vehicles that represents a production cost figure of $240,000,000.

There are a large number of additional uses for the Forklift Truck beyond those considered in this study. With slight modifications, it could be used to transport palletized loads. For example, construction firms need materials moved from one site to another. It normally requires a forklift on both sites and a truck as the transporting element between the two sites. This invention reduces vehicle requirements to one vehicle that conducts the whole operation. In another example palletized cargo could be transported from a warehouse to a department store, or from a supply storage facility to a customer. An extension of this type of operation could be that of a company in the transport and delivery business utilizing the Forklift Truck on radio dispatch. In fact, its uses are practically unlimited.

CONCLUSIONS

The Forklift Truck has tremendous potential. Design, fabrication, and testing of a full-scale prototype vehicle would require approximately two million dollars. System B would contribute to alleviating environmental problems associated with abandoned/wrecked vehicles on cost-effective bases (see Chapter 8). There are many other utilities of the vehicle not covered in this study. It could well be a Pareto Optimum resulting in net benefits for all parties (see Chapter 5).

10

Case Study in Traffic Management Systems: The Taiwan North-South Freeway Management Decision System

GEORGE CHEN, KENNETH KOU, AND HSI-I WANG

The three authors of this systems analysis were Republic of China citizens and members of my Systems Analysis class in Taipei, Taiwan during August and September 1977. Mr. George Chen and Mr. Kenneth Kou are operational managers in the Taiwan North-South Freeway project. Mr. Wang earned his B.S. degree in engineering and has a computer sciences background. The study illustrates:

1. Conceptualization of an automatic data-based system for operational decisionmaking with regard to a newly established automotive freeway system in Taiwan.
2. A systems approach to information needs for decisions with regard to traffic safety, toll-rate strategy, and toll station management and freeway operations.
3. Quantitative analysis using linear programming, stochastic models, probability theory, queueing models, Monte Carlo technique, and computer simulation.
4. A feasibility analysis addressing qualitative and quantitative benefits.

This study was an input to the planning process for the operations and control subsystem of the Taiwan North-South Freeway from Keelung to Kaohsiung.

R. M. KRONE

INTRODUCTION

The economic growth of the Republic of China's island province of Taiwan failed to reach 10 percent in only 3 of the 10 years from 1964 to 1973. Such a rapid growth is bound to cause some imbalance in different sectors of the economy. This has been the case with Taiwan's infrastructure. Its ports, inland transportation, international airports, power generation, and steel supply were rapidly becoming inadequate to meet the needs of the expanding economy.

Taiwan's present and future economic growth was considered along with a forecast of inland transportation demands. Regional land use patterns and development were defined and examined. The data gathered from these segments of the study when combined with the economic benefits, derived from having a freeway, proved conclusively that the project was feasible and desirable.

The Freeway, as shown in Figure 10.1, has a total length of 373 km, begins at Keelung in the North, and terminates at the Kaohsiung-Fengshan area in the South. The route passes through or near most of the major cities in the western plain of Taiwan. Besides directly connecting into the two major seaports at Keelung and Kaohsiung, there are access roads between it and the international airports at Taoyuan and Hsiokang. Closely aligned with the existing north-south arterial Highway No. 1, the Freeway is designed with a succession of moderate to flat grades and a strip of landscaped median dividing the northbound and southbound lanes. By definition, a freeway is a vehicle facility in which there is full control of access. That is, access is possible at interchanges only.

Planning for the freeway began in the mid-1960s and construction began in August of 1971. Completion of the construction phase of the Freeway occurred in 1979. This chapter addresses the area of operations management.

To meet the established objective of improving the overall transportation system in the western port of Taiwan, the Freeway should sustain an optimum level of service to keep traffic distribution in balance. Expansion of other transportation facilities and environmental changes in the western corridor can alter traffic distribution and affect the use of the Freeway. Hence the policy of operation and managing the freeway system has to be flexible enough to adapt to changing conditions. The scope of this flexibility includes varying toll rates and modifying certain physical features of the freeway. Both measures can be instrumental in maintaining its optimum level of service.

STATEMENT OF THE PROBLEM AND, OVERVIEW OF THE FREEWAY MANAGEMENT SYSTEM

A general management system of the freeway is constituted by road maintenance subsystems, management information subsystems, administrative and managing subsystems as well as traffic operating subsystems. So, the freeway traffic operating system we were dealing with is actually only a subsystem of the general management system.

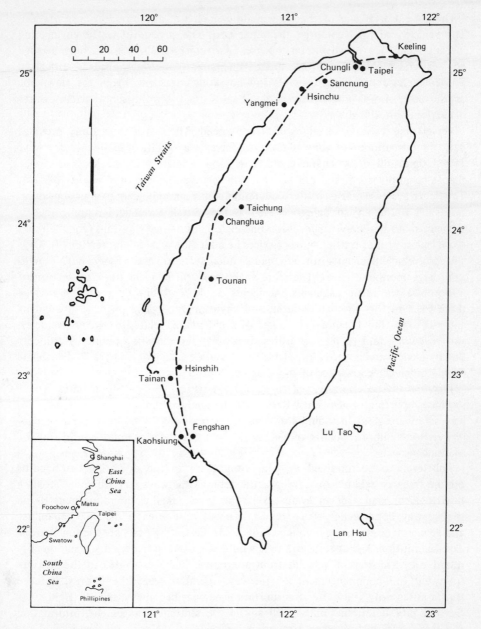

Figure 10.1 General location of North-South Freeway Project (Taiwan).

The road user is one of the three main elements of automobile transportation. The vehicle and road comprise the other two. The success of traffic engineering measures depends heavily upon the user. An understanding of not only average physical and mental limitations, but of the range of user response, is critical to proper exercise of traffic controls and operating measures. From the viewpoint of the user, travel time, safety and toll rates are the three significant measures of service of a toll highway.

Travel time varies inversely with travel speed. The travel time study provides data on the amount of time it takes to traverse a specified section of freeway. Travel time and delay characteristics are good indicators of the level of service that is being provided, and can be used as a relative measure of efficiency of flow. Traffic delays research is made to determine the amount, causes, location, duration, and frequency of delays as well as the overall travel and running speeds. Sequential studies allow for decision making on alternative solutions (e.g., "Should more lanes or other traffic control devices be added on a particular section?").

Motor vehicle accidents not only cause much suffering and misery, but, in addition, large economic losses. Therefore analysis of traffic accidents is of the utmost importance to traffic engineers. Accidents result from one or a combination of three basic factors: human, vehicle, and environmental. The purpose of accident analysis is to find the possible causes of accidents, as related to drivers, vehicles, and roadways, and to plan measures to protect the motoring public by reducing the frequency and severity of accidents. Accurate accident analysis is dependent on (1) thorough knowledge of the characteristics of drivers, vehicles, and roadway, (2) on the interrelationships of those characteristics, and (3) upon uniform and accurate reporting of accidents. Results of the analysis can lead to a rational freeway safety program including strategies for action, guidelines for programs, manpower needs, improvements of geometric (physical) design, and elimination of roadside hazards.

Toll levels are determined by comparing them to the operating cost-benefits on the freeway relative to travel via the arterial highways (see Chapter 7 for the discussion of standards in evaluation). The gross tangible operating benefits on the freeway come mainly from reduced transit time, greater distance capability, and economic gains through volume movement. Operating costs are the toll charges. The assumption is that toll rates that would wipe out user benefits would greatly retard the diversion of vehicles from alternative highway routes to the freeway following the commencement of freeway operation. In other words, induced traffic arises only if a better transportation bargain becomes available. Hence, for the toll rate determination as well as the toll station management, information and managerial techniques are required for systems analysis and decisionmaking. In fact, those three factors are interdependent and effected by other environmental factors such as road conditions and improvement of arterial highways. These components and considerations characterize a managerial system of the freeway traffic operation.

We feel this system should be constituted by two subsystems; the Data Bank Subsystem and Decision Subsystem as shown in Figure 10.2. Obviously, an enormous amount of information is involved. Neither the data handling nor the in-

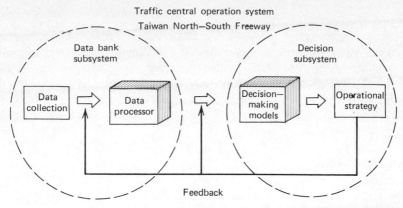

Figure 10.2 Traffic central operation system Taiwan North-South Freeway.

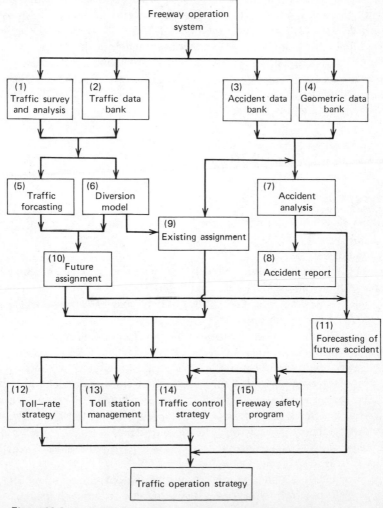

Figure 10.3 Logic chart for Taiwan N-S Freeway traffic operation system.

formation analyzing could be treated manually. A computerized management information subsystem and data bank subsystem should be established as components of the operating system. Models of systems analysis and decisionmaking should be studied and developed. Input and feedback will link the two subsystems. We think this kind of arrangement could maintain the viability of the traffic operation system. The linkages between the elements of this system are shown in Figure 10.3.

SCOPE OF WORK

As transportation problems exhibit characteristic system qualities of complexity and interdependence, the best solutions result from use of the systems approach (see Chapters 3 and 4). A study of the complete operating and management system of the freeway is beyond the scope of this paper. The specific three objectives we set for this study were (1) to establish the conceptual framework for a freeway operating system, (2) to outline a method for data collection and processing, and (3) to conceptualize a predictive and operational model for the decision process.

These objectives should meet the goals of providing more accurate, timely, and relevant information; providing a more efficient, effective, and economical decision process; providing the opportunity to measure the effectiveness of system implementation; and providing a safe and smooth operating freeway.

FREEWAY OPERATION DECISION MODELS

Decision Model on Traffic Control

One of the most important missions for freeway operation is to control the freeway and related highway system's traffic so that the freeway traffic operational efficiency can be maintained.

The objective of this decision model is to reduce the degree of congestion at bottlenecks (it is impossible to eliminate congestion completely), and to aid on-ramp vehicles merging into the freeway safely and effectively. Consequently, merging volume can be increased.

The North-South Freeway by-passes the old Taipei City but passes through the Taipei Metropolitan Area and acts as an effective toll-free urban freeway. Consequently, the intraurban traffic will seriously affect the freeway traffic. From western freeway operation experience, the following problems are expected: (1) traffic delay increases when traffic demand exceeds on-ramp capacity; (2) weaving traffic affects normal traffic and capacity; (3) as freeway traffic density increases, gaps between vehicles decrease and chances for on-ramp vehicles to merge decrease; and (4) congestion occurs when demand exceeds capacity in bottleneck areas.

These problems occur during peak hours. Peak-hour congestion might not be caused by deficient freeway capacity. It may be due to suddenly increased traffic on a certain ramp in a short period without diversion to other ramps. The capacity of different freeway sections may not be identical because of different geometric configurations. Hence, the basic concept of freeway traffic control is to distribute traffic to various sections in proportion to sectional capacities.

Since the upstream vehicles cannot be controlled by traffic signals, it is impossible to force them to leave the freeway before they reach their destination interchanges. However, the amount of traffic passing through the bottleneck areas can be determined from an origin-destination survey. Then the decision can be made to use either a message changeable sign to divert the traffic or use of ramp-metering to slow or stop the traffic from reaching the congested freeway segment. This requires a traffic control system capable of dispersing or diverting traffic when necessary. The problem then becomes data generation and interpretation to signal how much control is needed, where and how control should be applied, and for how long. The objective is one of balancing traffic flows throughout the system.

Linear programming is one means that can be used to help optimize traffic operations. The objective function can represent the output of the system at any given peak period. This should be maximized, subject to various constraints. Then if control measures are to be instituted they must prevent traffic congestion and decreased flow rate created by bottlenecks. To be effective, the mathematical model should depict how much traffic flow can be accommodated and the amount of flow that has to be diverted or stored until the critical period has passed.

In formulating a deterministic model for this process the volume of traffic at each input source can be combined to maximize the total output to the system subject to two types of constraints. First a set of constraint equations is required to assure that congestion will not develop at any location. A second set of constraint equations restrict the input from each source so that the demand at the entry ramps is not exceeded.

The objective function of this model is designed to maximize the total output. That is to maximize the Freeway volume to accommodate Freeway traffic to a desired level.

Maximize

$$\sum_{i=1}^{n} V_i \tag{1}$$

where

V_1 = volume at Sanchung entry ramp

V_2 = volume at Taipei entry ramp

V_3 = volume at Yuanshan entry ramp

V_4 = volume at Neihu entry ramp

V_5 = volume from Hsinchi Interchange

V_6 = volume from Keelung

From origin-destination data the percentage of each input volume can be obtained for each bottleneck location on the Freeway. The volume should not exceed the capacity of any single bottleneck. The entry volume should not exceed the demand for that ramp.

Subject to

$$\sum_{i=1}^{n} P_i V_i \leqslant C_i \qquad (2)$$
$$V_i \leqslant D_i \qquad (3)$$

where

P_i = probability of the ith entry volume over the section

V_i = volume of the ith entry ramp

C_i = capacity of the ith ramp

D_i = demand of the ith ramp

Having an acceptable balance stimulated using linear programming offers an opportunity to plan and implement a traffic control system to maintain that balance and alleviate traffic congestion. Computerized traffic signals, changeable message signs, and ramp metering are all successfully tested solutions to traffic control.

Safety Program Model

An effective safety program is essential for keeping freeway accidents to a minimum. Analysis of traffic accidents and trends is of the utmost importance to traffic engineers. Motor vehicle accidents cause human suffering and economic losses.

Accurate accident analysis is dependent on thorough knowledge of the characteristics of drivers, vehicles, and roadways, upon the linkages between those characteristics, and upon uniform and accurate reporting of accidents. Moreover, the purpose of accident analysis is to find the possible causes of accidents, as related to drivers, vehicles, and roadways, and to plan measures to protect the motoring public by reducing the frequency and severity of accidents. The numbers of traffic accidents and the consequently large amount of data make it impossible today to analyze accidents manually. Therefore, this operating system should employ an automatic data processing capability to provide an accurate and efficient accident recording system and make accident analysis easier through accident summaries and tabulations of related variables. Figure 10.4 diagrams the necessary components of the safety program data bank. Figure 10.5 shows the structure and data flows of the envisioned overall safety program system. Output documents

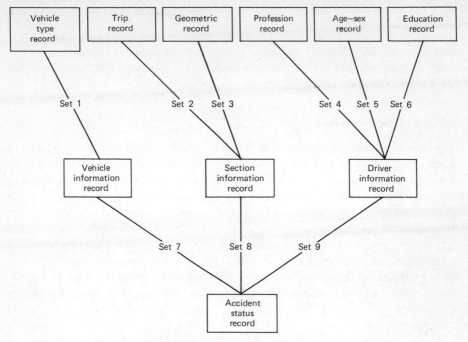

Figure 10.4 Safety program data bank.

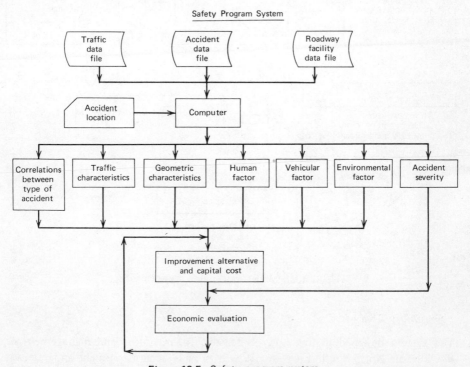

Figure 10.5 Safety program system.

of such a system could include (1) safety improvement alternatives; (2) capital cost for each improvement action; (3) reduction on accident cost after improvement alternatives; (4) decision on economical and effective alternatives; and (5) methods for measuring the effectiveness of implementation.

Toll Station Management

The primary criterion of managing a freeway toll place is the effective mix of numbers of toll collectors and procedures to provide rapid service to drivers. This requires finding the level of traffic delay that gives the best compromise between the conflicting objectives of economy and service. Figure 10.6 outlines the computer simulation queueing model to deal with this problem in quantitative terms. It enables the determination of the correlation among traffic volume number of toll booths and level of service. Once the cost of waiting is evaluated from the queueing model, the remainder of the analysis is conceptually straightforward. The management decision would be to determine the preferred level of service, which minimizes the expected cost of service and the expected cost of waiting for that service.

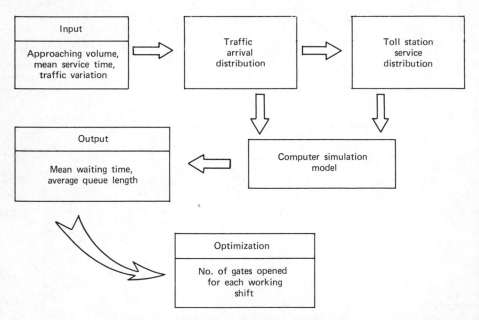

Figure 10.6 Model—toll station management.

Feasibility Analysis

The system designed in this study is feasible for operation and management of the Taiwan North-South Freeway. It will provide accurate, timely and relevant information for effective decisions in freeway operation. This study is designed to demonstrate the application of new technologies to leadership in the Freeway authority and the Ministry of Communications who have the decision responsibility for program implementation. To the driving public the program will provide enhanced transportation reliability, minimize human and material costs of accidents, and encourage public support. Those factors should produce both technological and political feasibility for this system.

Morever, economical feasibility can be evaluated through benefit/cost comparisons such as shown in Table 10.1. Time constraints for this study precludes detailed economic analysis.

Table 10.1 *Feasibility Analysis of Freeway Management and Operation System*

Operational Improvement	Benefit	
	Quantitative Benefit	Qualitative Benefit
Earlier improvement for release congestion	Saving on users cost	Provide adequate transportation reliability
Proper safety program to reduce accident	Decrease property damage	Minimize the society disaster
Proper management strategy	Saving on operating cost	Increase organization productivity & cohesiveness
Effective facility improvement program	Prevent over-investment for freeway system expansion	Enhance public favour Establish efficient budgeting program

Cost

Study and design cost

Systems construction cost { Establishment of traffic detecting facilities

Systems operating cost { Set-up computerize traffic operating and control center

CONCLUSIONS AND RECOMMENDATIONS

The authors make the following conclusions and recommendations as a result of this study:

Conclusions

1. The proposed system is theoretically valid and meets the three feasibilities (see Chapters 5 and 6).

2. The proposed system data bank would provide accurate information for freeway management and safety program future freeway development decisions.
3. Simulation of the models should occur before implementation in order to prescribe detailed data collection methods.

Recommendation

That the Taiwan North-South Freeway Authority adapt the traffic management and operations system proposal in this study.

11

Case Study in Nuclear Power: Systems-Evaluation and Quality Assurance: The Nuclear Power Plant

ROBERT KRONE

This application of the two-step evaluation procedure to the design phase of a nuclear power plant was developed as an illustrative and hypothetical example. I am indebted to Mr. James M. Dare, Consulting Engineer, Quality Assurance Services, International Division of the NUS Corporation for his assistance in the development of the case study. The case study illustrates:

1. The two-step systems analysis evaluation method provided in Chapter 7.
2. A medium for the systems analysis briefing (see Chapter 4) in the form of Table 11.2 which summarizes the evaluation process.

EVALUATION AND QUALITY ASSURANCE

The evaluation procedure, described in Chapter 7, involves two basic steps:

1. Determine the quality using criteria;
   Ask: "What quality exists?"
2. Judge that quality by comparison with standards.
   Ask: "How good is that quality?"

It is assumed that the nuclear power plant under evaluation is in the design phase. Research, development, and procurement decisions concerning major components of the power plant will be sensitive to the standards selected for evaluation. Required knowledge and expertise of quality assurance personnel is also sensitive to the standards selected.

The quality assurance program has a responsibility for the determination of the quality of the power plant systems. Traditionally it has accomplished this through a "go or no-go principle." That is, the quality assurance effort seeks to determine that minimum quality requirements have been met (i.e., the "go" criterion). Management can then assess whether quality achieved over the minimum is necessary for safety or is redundant and increasing costs with no additional benefits. The evaluation methodology of Chapter 7 provides a useful tool for both the quality assurance function and the management decisions. A typical quality program has three elements:

1. Quality estimation during the design and planning phase. This has been an engineering function with a goal of providing confidence that the final design and construction will assure the safety of the public;
2. Inspection and quality audits during the installation and construction phase. This is a quality assurance function to insure that government regulations and design criteria are met; and
3. Quality audits and inspection for compliance with the technical specifications throughout the operation phase to assure safety of the public during operation, maintenance, modification and refueling.

CRITERIA AND STANDARDS

The primary criterion of quality assurance is to provide confidence that a product will perform satisfactorily in service for a predetermined period of time. The secondary criteria are the qualities of the nuclear plant subsystems that are (1) reactor vessel, (2) reactor control systems, (3) emergency core cooling, (4) coolant recirculating system, (5) off-gas treatment, (6) fuel, (7) reactor intervals, (8) reactor structure, and (9) containment system. The actions necessary to achieve those qualities measured quantitatively are the focus of quality assurance and have been selected "for good reasons to be positively correlated with, and more measureable than, the primary criterion" (see Chapter 7 for elaboration of the theory of systems evaluation).

Three illustrative standards were selected for this case study and are listed in Table 11.1 as Options A, B, and C.

A summary of the results of the two-step evaluation process for this hypothetical study is displayed in Table 11.2. The quality determined for Step 1 of the evaluation process is shown in columns b, c, and d. Note that the standard selected for Step 2 of the evaluation process has a bearing on the method of presenting the results determined in Step 1. For Option A (government regulations) the quality determined is expressed in terms of percentage of prescribed safety standards met, whereas in Options B and C the quality determined (columns c and d) is expressed in terms of reliability criteria met for 20 or 40 years plant life span, respectively. Columns e, f, and g show the results of Step 2 of the evaluation process and reflect the judgment of how good the quality determined in Step 1 is in comparison with the standard selected. That measurement presentation uses an absolute scale—that is, judgments of "too good," "satisfactory," "marginal," and "unsatisfactory" are made as a result of the quality assurance program eval-

Table 11.1 *Three Evaluation Standards for a Nuclear Power Plant*

Option	Standard
A	Government Regulations (Safety Standards)
B	Short Plant Life Span* (Reliability for 20 years)
C	Long Plant Life Span* (Reliability for 40 years)

*Nuclear Power plants to date have been designed for a 40 year plant life, however, consideration of a standard for a shorter life span would be an appropriate systems analysis in itself.

uation. Comments on those judgments follow. The Roman numerals are cross-referenced to column h of Table 11.2.

I. The reactor vessel would probably be designed for a 40-year life span regardless of option selected.

II. The reactor control system must be of very high quality for any option.

III. Assuming the 135% quality determined in Step 1 (government standards) and shown in column b, this could be judged "too good" for Options A and B but satisfactory for Option C.

IV. With this quality determined in Step 1 the selection of evaluation standard will have a direct influence on design, procurement, and construction as quality judged as "too good" for government standards (Option A) and for a 20-year plant life span (Option B) might be judged as "unsatisfactory" for a 40-year plant life span standard (Option C).

V. Planning for component replacement of the Off-Gas Treatment System will be different with Options B and C. Management should decide, a priori whether the government standards will provide the reliability desired. Construction costs will be higher if Option C is desired but replacement costs may be even higher than initial construction costs if a 40-year life span is expected and the necessary quality has not been designed into the system.

VI. Fuel replacement schedules and costs will probably not be affected by the choice of these three options.

VII. Reactor internals are usually overdesigned for safety reasons so they would not vary with option selection.

VIII. This is another example of decision sensitivity to options and standards selected. Management judgment concerning the 150% quality determined under government standards will be influenced by further selection of short or long plant life span (i.e., Options B or C).

IX. If, in the design phase this quality of 85% of government standards were determined by Step 1 of the evaluation system, management action would definitely be required to bring the containment system up to an acceptable quality for Options A and to even higher quality if Option C is also picked.

Table 11.2 *Example of the Application of the Systems Analysis Evaluation Procedure to a Nuclear Power Plant: Focusing on Quality Assurance (All Figures and Judgments Hypothetical)*

Primary criterion: To provide confidence that the plant will perform satisfactorily for a predetermined period of time

Columns: a	b	c	d	e	f	g	h
	Evaluation System Steps						
	Step 1: Quality Determined Standards Met, %			Step 2: Judge that Quality Comparing with Standards			Comments (see text)
Secondary criteria; Qualities of plant subsystems	Option A	Option B	Option C	Option A Government Standards	Option B 20 yr	Option C 40 yr	
1. Reactor Vessel	105	125	100	Satisfactory	Satisfactory	Satisfactory	I
2. Reactor Control System	110	125	100	Satisfactory	Satisfactory	Satisfactory	II
3. Emergency Core Cooling	135	150	100	Too good	Too good	Satisfactory	III
4. Coolant Recirculating	120	120	75	Too good	Too good	Unsatisfactory	IV
5. Off-Gas Treatment	100	97	88	Satisfactory	Satisfactory	Marginal	V
6. Fuel	110	120	105	Satisfactory	Satisfactory	Satisfactory	VI
7. Reactor Internals	125	110	100	Too good	Satisfactory	Satisfactory	VII
8. Reactor Structure	150	103	94	Too good	Satisfactory	Marginal	VIII
9. Containment System	85	100	50	Unsatisfactory	Satisfactory	Unsatisfactory	IX

*Measurement scale: Too good = unnecessary resources expended; Satisfactory = quality judged satisfactory; Marginal = quality judged marginally satisfactory, improvements probably needed in future; Unsatisfactory = quality is unsatisfactory, improvements or redesign needed.

CONCLUDING COMMENTS

In this illustration of the two-step evaluation process of Chapter 7, nuclear power plant decisionmakers could be provided, during the project design phase, with useful data on which to make critical decisions and planning guidelines for the subsequent phases of construction, testing, operations, and deactivation.

The method of summarizing the data, shown in Table 11.2, with the attached comments is an example of an appropriate medium for the systems analyst's briefing of the results of the evaluation (see Chapter 4 for a discussion of the systems analysis briefing). All data used are hypothetical to demonstrate the applicability of the evaluation model. My selection of standards is oversimplified. Some more complex mixture of rational and extrarational standards would actually be used but the methodology would be the same.

In systems with complex technology and high failure impact, design phase evaluation becomes one of the most critical responsibilities of management. In the case of nuclear power plants the March 28, 1979 failure of the unit 2 coolant system at the Three Mile Island Nuclear Power Plant near Harrisburg, Pa. suddenly produced renewed interest and efforts toward more effective evaluation of nuclear plants as well as more evaluation and analysis of the alternatives to nuclear energy generation systems. At the Three Mile Island Plant there were fears that an unprecedented core melt-down might occur that would release radioactivity of dangerous levels. Unforeseen and unfortunate events will occur in spite of the best evaluation efforts, but management can reduce both the probabilities and magnitude of those events through continual review and validation of the secondary criteria being used and the standards being applied in the evaluation process.

12

Case Study in Technology Transfer: Management Knowledge for Technology Transfer

ROBERT KRONE

Conceptualizing appropriate analysis models for complex problems is an essential macro tool of the systems analyst. International technology transfer presents a complex business, legal, social, and political problem for both the technology provider and receiver. This chapter provides conceptualization and structure for the identification of management knowledge prerequisites for technology transfer. In so doing it creates a dual-purposed model—first it is a behavioral model of collective knowledge currently in use; secondly, it is a normative model for what knowledge should be used by managers and decisionmakers involved in technology transfer. The objective is to increase the probabilities that technology transfer programs will meet the criteria of applicability, effectiveness, efficiency, and feasibility. If management can, in the policymaking and planning stages, acquire relevant knowledge the probabilities of success will increase and the judgment of the desirability of pursuing negotiations can be made with higher confidence. The model emerges from past and current management knowledge about technology transfer actions to provide the complex structure of knowledge needed for future technology transfer ventures. It is presented in a taxonomy form at the macro level. Space constraints preclude expanded discussion of the knowledge prerequisites or the inclusion of concrete examples.

For simplification I will use the terms and associated acronyms "transnational corporation (TNC)" and "host nation (HN)" as representative titles for any two parties involved in the technology transfer process. I recognize that the entities involved vary widely across public and private spheres. The model, as a whole, is not sensitive to substitution of specific entities for the terms transnational corporation and host nation although some of the discrete knowledge prerequisites will be found more appropriate than others once those entities have been selected.

An earlier draft of this paper was presented at "The International Confer-
ence on Management Implications of the Transfer of Technology from the
Developed to Developing Countries" jointly sponsored by the Academy of
International Business (ATB), The United Nations International Development
Organization (UNIDO), and the Federation of Korean Industries. The Conference held in Seoul, Republic of Korea, June 12–17, 1978 was the first
international conference to address this theme. I am indebted to Dr. William
Bredo for his suggestions, which resulted in improvement of that earlier
draft.

A successful technology transfer is one that is conceived, designed, planned, and
implemented to meet, from the viewpoints of both technology provider and technology provider and technology receiver, four secondary criteria (see Chapter 7):

1. *Applicability*. Relevance to values, goals, and needs of both parties—the transnational corporation (TNC) and the host nation (HN).
2. *Effectiveness*. The degree to which the technology transfer program meets
   its goals.
3. *Efficiency*. The degree to which resource utilization is optimized.
4. *Feasibility*. The probability that economic, technological, and political feasibilities of the technology transfer venture will be met (see Chapters 5 and 6).

If confidence in achieving the above criteria can be met, then the technology
transfer program is desirable and should be aggressively pursued or continued.
This paper takes a macro approach to identify technology transfer knowledge
requirements to meet the four criteria. These four criteria categories are not mutually exclusive but, rather, are mutually supportive. Where a knowledge prerequisite overlaps into two or more categories I have made an arbitrary choice to
avoid redundancy.

1. APPLICABILITY: RELEVANCE TO VALUES, GOALS AND NEEDS OF BOTH PARTIES

A. Policy Sciences (see Chapter 5) knowledge to include:
   (1) Values analysis to identify areas of values consonance between the TNC
and the HN. Values analysis is perhaps the most difficult knowledge requirement
to satisfactorily obtain, but it is essential to identify potential values conflicts
and to answer the critical questions of "technology transfer for what?" and "technology transfer for whom?"
   (2) Policymaking system knowledge to understand the infrastructure for
technology transfer to identify where, how, and by whom decisions are made,
what structural units exist for short- and long-range planning, and how the technological needs of a nation or locality are determined.
   (3) Policy strategies. Policy strategies will be strongly linked to values and

will not all be explicitly stated, however strategies toward risk, toward short- or long-term time preferences, toward present or future discounting, and toward the degree of managerial knowledge and technological knowledge absorption that is desired by the HN and the TNC and over what time period should be known. Failure to explicate these strategic preferences in technology transfer contracts will inevitably lead to frustration, misunderstanding, and even venture failure.*

(4) Policy analysis methods, institutions, and policy analysts involved in technology transfer. An atmosphere encouraging interaction between the relevant policy analysts and systems analysts of the TNC and HN will help build the knowledge requirements cited herein.

B. Environmental knowledge to determine the level of technology is, or should be, applicable (e.g., traditional, intermediate, appropriate, or scientific-forefront technology) and for what period of time contracts should be negotiated:

(1) This type of knowledge includes portions of all the four knowledge criteria categories of applicability, effectiveness, efficiency, and feasibility.

(2) Knowledge of the spectrum from simple traditional agricultural technology in very low GNP/capita countries as host nations to complex turn-key technology transfers such as computer systems, nuclear power plants, or fertilizer plants.

(3) Cross-cultural knowledge of the TNC and HN personnel.

C. Host nation industrial development knowledge and marketing knowledge to determine the appropriate economic sector for technology transfer (e.g., heavy industry, pharmaceuticals, agriculture, defense, textiles, health, transportation, communications, or energy).

D. Formal constraints and incentives compose the investment climate:

(1) HN strategy toward economic development plus knowledge of HN and TNC to meet economic and industrial goals. Implicit in this knowledge requirement is the awareness of the trends in political stability, or the effects of cyclical political activity (e.g., the United States four-year presidential election cycle).

(2) Business laws and regulations, and their trends, relating to foreign investment, for instance, knowledge of trade barriers; taxes on dividends; effective rate of tax on earnings; convertibility of currency; availability of hard currency; labor availability, costs, and status of labor movements; required labor agreements; size of market; distribution system and competition; and protective patent and copyright laws for intellectual products. This assumes a capability of current analysis and future studies.

---

*This point was illustrated by one of the case studies submitted to the Seoul 1978 Conference by Professor Ku-Hyun Jung and Professor Kee-Young Kim, College of Business and Management, Yonsei University, in their paper, "Transfer of Technology and Management Knowhow through Multinational Corporation: A Case Approach." In their "Case 4: A Petroleum Refinery" pp. 12–13, they describe how in one of the first and biggest petroleum joint ventures in Korea failure to explicate some of these type policy strategies resulted in Korean dissatisfaction with United States expatriate management when managerial transfer to the Koreans did not keep pace with Korean absorption of technological skills.

(3) Rigidity or flexibility in the planning and implementation process. This should include cross-feed knowledge within the HN of intragovernmental agency actions.

E. Profit expectations to include differences in views between the TNC and the HN and knowledge of those profit expectations obtainable in the domestic market.

F. Mutual TNC and HN knowledge and acceptance of the probable cultural, human, and ecological impacts during the technology transfer and afterwards as a result of the diffusion of the technology. This includes the impact on the preservation or disruption of traditional culture; the relationship to nationalistic development attitudes of the HN; the phenomena of potential elimination of a HN industry as a result of technological substitution.*

G. The transfer of technology often leads to changed organizational behavior, which may impact on independent, but interfacing, businesses within the HN such as suppliers, distributors, and competitors.

H. Research and development knowledge impacting on the technology transfer venture under consideration. The capability of a developing HN to acquire this knowledge prerequisite is limited. The TNC may have a distinct advantage.

I. Knowledge of technology transfer failure and disinvestment cases.

J. My final entry under the criteria category of applicability is that of general system knowledge. The system approach (see Chapter 3) is important in meeting technology transfer knowledge criteria to accomodate widely varying perspectives and to answer the question: "Is it applicable?" Dr. H. W. Pack, from UNIDO, in Vienna, during his presentation at the Seoul 1978 conference, provided participants with his diagram of "The Technology Transfer Elephant," which showed four blind men feeling different portions of the elephant's anatomy and making diverse conclusions. The components of "The Technological Transfer Elephant" were: Management, technology policy and plan, investment promotion, flow of information, technological advisory services, appropriate choice of technology, technological capabilities, social and economic constraints, R & D, design and consultancy, pilot plants, full-scale testing, standardization, quality control and pollution.

2. EFFECTIVENESS OF TECHNOLOGY TRANSFER—
THE DEGREE TO WHICH THE TECHNOLOGY
TRANSFER PROGRAM MEETS ITS GOALS

A. Knowledge for conceptualizing effectiveness criteria. This includes the establishment of TNC and HN mutually agreed secondary criteria (see Chapter 7). The identification and measurement of valid secondary criteria is probably the one most important knowledge prerequisite to insure effectiveness.

B. Knowledge of the TNC and the HN respective views on innovation and the role of entrepreneurship, and on the preferred mix of expatriate (TNC) man-

*Derek Medford, the Director of the Center for Applied Studies at the University of South Pacific, Fiji described to the Seoul 1978 Conference how this has occurred with the sugar industry of Fiji.

agement to HM management at all levels. The issue of management growth potential for employees is linked to this determination.

C. Knowledge of the relationship of management knowledge to past effectiveness to include possibilities for improvement over time. This assumes familiarity with available technology data, case studies, and current issues (e.g., the technology transfer international code of conduct issue).

D. Explicit knowledge on which to make the decision of which management functions remain with the TNC, which will be assumed by the HN and which will be areas for training for the HN personnel. These functions include: technology transfer design and strategy, policymaking, planning, goal setting, organizing, budgeting, programming, motivating, innovating, evaluating, scheduling, controlling, supervising, leading, communicating, developing, forecasting, and monitoring of operations. Management function transfer will vary with the type of contract (e.g., wholly owned subsidiary, joint venture, licensed subsidiary, management contracts, or direct investment). This is also an area for management evaluation during subsequent phases of technology transfer. Alternative models of joint management for increased effectiveness should be a periodic agenda item for top management.

E. Knowledge of technological forecasting institutions, units, and techniques, as well as plans of the TNC and the HN for their utilization. National governments are rapidly improving the knowledge base for this prerequisite.

F. Knowledge of what institutional relationships have been developed by the TNC or HN to assist in the acquisition, processing, and analysis of technology transfer information to include management consultation, use of published works or technology transfer information systems. There was in 1979 no international data bank for a HN or TNC to turn to in its considerations of potential technology transfer ventures. The Soviet Union has one which, reportedly, tells who has what technology and what has been sold at what price. Multinational corporations and large national corporations all maintain their own privileged sources of information related to technology in their areas of interest.

G. Knowledge of perceptions of business ethics—both within the TNC and HN.

H. Knowledge of current international efforts to regulate technology transfer.

I. Knowledge requirements for TNC agents of technology transfer.

J. Knowledge to maintain HN government and TNC negotiations and dialogue after the technology transfer agreement has been signed.

3. EFFICIENCY—THE DEGREE TO WHICH RESOURCE UTILIZATION IS OPTIMIZED

A. Knowledge of specific management control techniques to be shared between the TNC and HN (e.g., production scheduling techniques, budgeting, cost control, quality control, inventory control, reports generation and distribution, personnel management, to include promotion standards and evaluation techniques, and labor relations mechanisms).

B. Production knowledge gained through productivity indices such as cost/unit, units/time, or malfunction trends.

C. Personnel training knowledge to improve human skills to deal professionally with technology transfer.

D. Labor availability and costs.

E. Material resources location, reserve levels, costs to obtain, reliability of supply over time, transportation and distribution considerations.

F. Knowledge of energy conservation methods, waste utilization technology and areas of potential cost reductions through capital investment in conservation technology.

G. Crossfeed knowledge of other similar technology transfer operations.

4. FEASIBILITY—THE PROBABILITIES THAT ECONOMIC, TECHNOLOGICAL, AND POLITICAL FEASIBILITY REQUIREMENTS OF THE TECHNOLOGY TRANSFER PROJECT WILL BE MET

A. Economic feasibility is the probability that resources will be available for the technology transfer venture. This is fundamental to business and needs no further elaboration.

B. Technological feasibility is the probability that the technological and scientific goals for the technology transfer project will be met. This is primarily the responsibility of the TNC but knowledge for determining technological feasibility of the TNC proposals is essential for the host nation. If a HN is interested in scientific forefront technology both the risks and potential benefits will be greater. There is no substitute for scientific and technological expertise and experience for meeting this criterion.

C. Political feasibility is the probability that a technology transfer proposal will be acceptable to the relevant decisionmakers. Political feasibility, political power and consensus building needs operate at every level of private and public organizations. It is a normal phenomenon of public and private life which cannot be excluded from the knowledge criteria for successful technology transfer. The ability to demonstrate convincingly that both economic and technological feasibility exist is a necessary, but not a sufficient, prerequisite to achieve political feasibility. Political feasibility is dependent upon a large number of extrarational variables in the environment as well as the internal organizations of the two negotiating parties. Completely ignoring political feasibility will result in missed opportunities, wasted resources, misunderstanding, values conflicts, and even international incidents. On the other hand, failing to offer innovative proposals that may seem to be outside the current domain of political feasibility, but nevertheless judged by proponents to be advantageous to the TNC and HN, can also result in lost opportunities for both parties.

The random impacts on political feasibility for a technology transfer venture stemming from international politics cannot be predicted. The blockage by the government of the United States of the sale of a Sperry Univac computer system to the Soviet Union in July 1978 after the conviction of dissidents Anatoly Shcharansky and Alexander Ginsburg of crimes against the Soviet Union is an illustrative example of a random change in political feasibility. The knowledge by TNCs that such random actions can and do occur plays a role in their overall judgment as to applicability of any technology transfer venture under consideration.

EPILOGUE

The problem of international technology transfer presents the most complex set of variables in the business world. Fully meeting all of the knowledge requirements outlined above would rarely be feasible for either party involved in the technology transfer venture. Grossly inadequate or erroneous knowledge in any of the four criteria categories of applicability, effectiveness, efficiency, or feasibility will doom the project to failure from the beginning. Even marginal improvement in management knowledge for technology transfer in the four criteria categories can mean the difference between failure and success.

13

Case Study
in Agricultural Systems:
The Spencer Soil-Penetrating
Phosphate Discovery

V. E. SPENCER

The Spencer Case Study is in two sections. The first section, titled "How to Produce More Food for a Hungry World," was a paper prepared for review by participants of the Limits to Growth 1975 Conference, The Woodlands, Texas, October 19–21, 1975. The second section titled, " Birth of a Better Fertilizer," is an essay relating the circumstances, commencing in June 1928, which led to V. E. Spencer's discovery of a soil-penetrating type of phosphate. The Spencer Case Study illustrates:

1. Behavioral and normative research within the scientific method.
2. The functioning of creativity and tacit knowledge in scientific discovery.
3. Research resulting in the establishment of technological feasibility for a system improvement (i.e., enchanced plant nutrition) but which has yet to achieve political feasibility for further research to determine economic feasibility.

V. E. Spencer was born November 17, 1893. He earned his B.S. degree in Agriculture in 1915 and his M.S. in Chemistry, with a major in Organic Chemistry, in 1926. He dedicated his professional life to the study of chemistry of the soil and plant nutrition research with the majority of his field experiments being conducted in the State of Nevada during his tenure as Professor of Agriculture and Soils Research Chemist at the University of Nevada, Reno, from 1928 to 1960. He is today the world's authority on the capability and potentials of increased crop production through the application of soil penetrating phosphates.

R. M. KRONE

HOW TO PRODUCE MORE FOOD
FOR A HUNGRY WORLD

Most of the food eaten by all the world's people comes from the soil. That has always been true, and it will continue to be so in the foreseeable future.

Now, and for several years past, genuine concern is being expressed, and with increasing frequency, that the world's food supply may fall short of minimum needs in the present decade, or seems certain to do so in the next. In fact, we are told that there are sizable areas of the world where hunger, malnutrition, and even starvation are plaguing millions of people right now.

Since the soil is the primary source of the world's food supply, and the time will soon be here when that supply will be inadequate, it naturally follows that something is needed to make the soil much more productive. Customarily, man has met that need with fertilizers, and for many years they have filled the bill nicely. But the good effect of fertilizers in increasing crop yields appears to be on the wane. Some evidence of this, presented in an article* by Dr. James G. Horsfall, until recently director of the Connecticut Agricultural Experiment Station, indicates that for many crops, fertilizers are no longer increasing yields significantly.

Many years ago—in late June, 1928, it was—I discovered a type of phosphate that will permeate a soil mass many times deeper (thicker) than the plowed layer, under the influence of percolating water. That had never been done before, neither here in the United States nor anywhere abroad. Up to that time, such a phosphate had never been used or proposed for use as a fertilizer.

As every soils specialist doubtless knows, no phosphatic fertilizer now used, or that has ever been used, penetrates downward into the soil to any extent, because the soil quickly "fixes" it at the spot where it is applied. I could cite numerous items of evidence to that effect, but the following statement in a letter received by a colleague from Dr. H. Rex Thomas, Acting Associate Administrator in the Agricultural Research Service, U.S.D.A. in January, 1971, will suffice: ". . . the great mass of research evidence shows very conclusively that phosphate applied to soils does not move downward except in very minute amounts. Liberal use of phosphate fertilizers has resulted in accumulations of phosphorus in the surface soil."

The main circumstance that induced me to seek and find a soil-penetrating phosphate was the failure of superphosphate treatments, applied in an experimental orchard in Southern Nevada in 1920, to produce any beneficial effect.

I reasoned that the lack of response to the phosphatic treatment probably was due to failure of the phosphate to penetrate downward sufficiently to contact enough of the trees' root systems (that is, to contact enough of any single tree's root system) to have any appreciable effect on the fruit production.

In addition to failure of applied phosphate to penetrate downward to any extent, which failure deprives most of a root system of the phosphate meant for

---

*"Portents . . . and Aloha." In *Frontiers of Plant Science,* November, 1971; published by the Connecticut Agricultural Experiment Station.

it, there may be another adverse effect, on plant growth, due to that lack of penetration. It is possible that accumulation of much phosphate in the plowed layer of soil, where treatment with phosphatic fertilizer has been frequent and long continued, can contribute to a lessening of plant growth. Presumably it isn't that the accumulated phosphate would become toxic or harmful per se, to the plant, but that the sharp difference between the high concentration of phosphate in the layer of accumulation (plowed layer) and the much lower concentration of phosphate in the soil below that layer, would operate to slow down the growth of the plant.

This speculation regarding a possible ill effect on plant growth due to a high accumulation of phosphate in a small portion of a root zone is not without foundation. I have conducted experiments involving what I have termed "differential feeding of a single root system" and in them have obtained some amazing results. By "differential feeding of a single root system" I mean feeding different portions of a single root system different plant nutrients, respectively. The results I got in experiments of this type were an additional compelling reason why I felt it essential to achieve a soil-penetrating phosphatic fertilizer.

Some details regarding the discovery of a type of phosphate capable of penetrating soils are related in an article by V. E. Spencer and Robert Stewart, entitled "Phosphate Studies: 1. Soil Penetration of Some Organic and Inorganic Phosphates," published in *Soil Science,* Vol. 38, No. 1, July 1934.

Two field experiments were made involving comparisons of organic, soil-penetrating phosphate with the conventional inorganic, nonpenetrating form. The results from the first of these are presented in Figure 13.1. In that experiment the penetrating form gave a good account of itself, even though the rate of phosphate application was much too low for best effectiveness. Of course, the severe drought greatly limited the forage yields in the last two years of the experiment.

The second experiment was made on a ranch in Nevada and with a different penetrating organophosphate, with alfalfa the crop. In this experiment the total amount of phosphate applied, per acre, was equivalent to 450 pounds of P_2O_5, which was seven times the rate used in the first experiment. Also, the plots were under irrigation, so there was no drought to curtail the crop growth. These circumstances gave the phosphates a much better opportunity to boost yields than they had in the first experiment.

As the second experiment progressed, naturally I had a surge of satisfaction at the way it was going. The penetrating organic phosphate pulled ahead strikingly and was on the up grade, so I was most anxious to continue the experiment and learn how high it would go. But the director of the Nevada station abruptly and (to me) unexpectedly terminated the project under which I was doing that work, so the 1945 results were the last I got. My guess is that the organophosphate would have climbed up to a production several times that of the superphosphate.

It must be kept in mind that the deeper the mass of soil permeated by a soil-penetrating phosphatic fertilizer, the greater the amount of that phosphate which must be applied to ensure that a high-enough concentration of phosphate exists at any and all points within the soil mass to be invaded by the foraging crop roots. Application of enough phosphatic fertilizer to a sufficiently deep soil layer to bring it into that desirable condition will usually require several years; for in the main, growers will not be able to afford applying that much phosphate all at one time. However, the fertilization will be profitable to them as it progresses, so it will impose no financial burden.

Figures 13.1 and 13.2 indicate that there is a vast storehouse of food below the plowline, the world over. Not just for wheat and rice, but for all crops. That applies to every acre of land that has been, is, and ever will be brought under cultivation. For simple, incontestable reasons, deeply placed phosphate is the only key to unlock that storehouse. I applaud the "Green Revolution," but it furnishes humanity no such key!

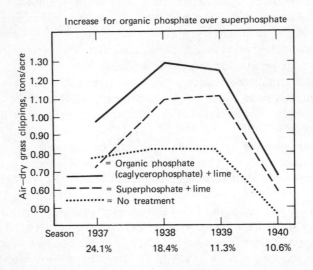

Figure 13.1 Comparative performance of a soil-penetrating organic phosphate and a non-penetrating inorganic phosphate as fertilizer on pasture sod. These tests were made by Dr. Howard B. Sprague, Agronomist, New Jersey Agriculture Experiment Station. The plots were located on Voorhess Meadow, an experimental area operated by the New Jersey station.

The organic, soil-penetrating phosphate was calcium glycerophosphate. It was supplied to Dr. Sprague by the Soils Department, Nevada Agriculture Experiment Station. The inorganic, non-penetrating phosphate was superphosphate. Each phosphate was applied, as top-dressing, at the rate of 64 pounds of P_2O_5 per acre, in the spring of 1937.

The increases for organic phosphate over superphosphate in 1937 and 1939 were statistically significant beyond the 5 per cent level, and the increase in 1938 was significant beyond the 1 per cent level.

A severe drought in 1939 killed practically all the white clover, and clover did not come in spontaneously in 1940. Regarding that, Dr. Sprague wrote: "There is very little response to mineral fertilizers alone, in the absence of a legume. Under the circumstances I do not feel that 1940 was at all a fair test of the value of the organic phosphate." Later Dr. Sprague commented further, as follows: "The low yields of all plots in 1940 have been explained as due to drought." and: "There is no question in my mind that calcium glycerophosphate was superior to superphosphate on Voorhees Meadow." and: "The returns in mixed grass and clover herbage for 1937, 1938, and 1939 indicate that availability was much greater for the organic phosphate."

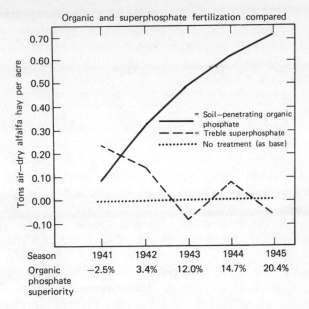

Figure 13.2 caption table:

Season	1941	1942	1943	1944	1945
Organic phosphate superiority	−2.5%	3.4%	12.0%	14.7%	20.4%

Figure 13.2 Above percentages show the superiority of organic phosphate over superphosphate, on the basis of treatment. The organic phosphate was calcium ethyl phosphate. Each phosphate was applied at the rate of 225 pounds of P_2O_5 per acre. The experiment was made on the Dangberg ranch near Minden, Nevada. In August 1940, just prior to planting the alfalfa, the two phosphates were broadcast and then plowed in with a disk plow. The same applications were repeated in April, 1943, as top-dressings on the established alfalfa.

POSTSCRIPT ON "BIRTH OF A BETTER FERTILIZER"

Here I relate the circumstances that led up to my discovery of a soil-penetrating type of phosphate. To do this, I should go back to the year 1928, when I came to the University of Nevada from the University of Illinois. I reported for duty here at Nevada on April 1, 1928. I was employed in the Nevada Agricultural Experiment Station. My immediate superior in that work was Dr. Robert Stewart, who was a member of the experiment station staff and was also the dean of the College of Agriculture here in the University of Nevada.

Prior to my coming here Dean Stewart had established a soil fertility experiment field in the vicinity of Las Vegas, Nevada. Various fertilizers and other soil treatments were being tried out on that field. In addition to those trials there were also some cooperative experiments in some orchards in the Las Vegas Valley. I'll refer to those orchard trials again, a little later.

At the time there was an organization known as the Western Society of Soil Science. This society held a series of meetings annually. In that year 1928 its meetings were held in Riverside, California. Dean Stewart felt that I should attend those meetings, so I did. A good many research papers were presented at the meetings; that is, reports on experimental work that had been done in the several western states.

Among those many reports, one in particular impressed me greatly. It was a report about trouble that had developed in prune orchards in California. The research people who were working on that problem had given it a name, "prune die-back," and that name indicates the nature of the trouble. The small outer branches of the trees would shrivel and die, beginning at the tips and gradually progressing inward to the thicker parts of the branches. Thus, eventually the whole tree was killed.

The research men who were working on that problem had found what they considered to be the cause of the die-back. They were convinced that the soil in the orchards where the trouble occurred was not supplying sufficient potash to the trees; that is, the nutritive element potassium. Once they had found that out, they thought they had the problem licked. All they would have to do, they supposed, was to make a liberal application of potash fertilizer to soils in orchards where the trouble was showing. So, their next step was to apply potash fertilizer in some areas where the trees were affected. They expected the affected trees to show improvement.

Naturally they were greatly disappointed when it became apparent that there was no improvement. And their first thought, then, was that after all their theory about the trees' need for potash was not correct. However, one of them was not satisfied on the latter point. He suggested that they sample the soil inch by inch, from the surface downward, and analyze those samples for potash. So that was done. What they found was very interesting. They found that a surface layer of soil had been greatly enriched with potassium, but that the layer was only five inches thick. In other words, although a heavy application of the potash fertilizer had been made, none of it had penetrated downward more than five inches, and the penetration to that depth had not helped the trees any.

After learning that, they found a way to get the potash fertilizer down into the deeper soil, and thus into contact with a large portion of the trees' roots. They accomplished that by boring a large number of holes down deep into the soil under and around the trees, and filling the holes with the fertilizer. The irrigation water dissolved the fertilizer and caused it to seep a little into the surrounding soil, where it was fixed; that is, prevented from moving further. The trees' roots could then contact enough of the fertilizer to do some good. The result was decided benefit to the trees. In a published report on that work, the California researchers stated that the method they used for applying the fertilizer is not applicable on the large (orchard) scale, but had proved adequate for the experimental use.

This experience with the potash fertilizer was certainly an eye-opener to me. I had never thought of the effectiveness of a fertilizer as being in any way dependent on the extent to which the fertilizer penetrates down into the soil, and I am sure the same has been true of practically all soils specialists and agronomists the world over. In the main, apparently the assumption has always been made—doubtless unconsciously—that any fertilizer put onto the surface of the soil, or into a shallow surface layer, would contact all the crop roots for which it was intended, and thereby benefit the crop if the plants needed that kind of fertilizer. Yet that work on the prune die-back problem gave a beautiful illustration of the need for penetration of soil by fertilizer.

After the meetings at Riverside I took a side trip to Las Vegas, Nevada before returning to Reno. I wanted to become familiar with the soil fertility experiments in progress there near Las Vegas, and also with the fertilizer trials in some orchards in that area. Mr. George Hardman was in charge of all that experimental work. He told me that he was disappointed in the results of the orchard trials, because none of the fertilizer treatments had had any beneficial effect on the fruit production. When he said that, my mind jumped again to the findings of the California researchers who had worked on the prune die-back problem; that is, the failure of the potash fertilizer to penetrate deeper than five inches into the soil, and the consequent lack of benefit to the prune trees.

The circumstances encountered in the California prune orchards, and in the orchards in southern Nevada, were strikingly similar. In both cases fertilizer had been applied to the land, but had had no beneficial effect. In the case of the prune orchards, the failure was due to lack of penetration of the soil by the fertilizer; that had been proven. Therefore, I concluded that the ineffectiveness of the fertilizer on the Nevada orchards was due to the same cause; namely, lack of soil penetration.

On my return to Reno from Las Vegas, I told Dean Stewart about the report of the research work done in California on the prune die-back problem, and also about the failure of fertilizers to benefit orchards in southern Nevada. He concurred in my opinion that the reason the fertilizers had not been effective in those orchards was that they had not penetrated downward through the soil to any extent, and thus had not contacted enough of the trees' roots to benefit the trees.

A day or two later he and I were sitting by my desk in the soils research laboratory mulling over the situation. Suddenly the dean spoke up—rather plaintively. His exact words were (I quote): "If only there was some way to get phosphate down to those tree roots!" Instantly I spoke, and heard myself saying "I think I know a way to do that." Later, looking back at that incident, I was rather amazed that I had said those words, for I had not had them in mind to say. I had not been trying to think of a way to cause phosphate to penetrate down through the soil; at least not consciously. What had happened was that the instant the dean made the statement, a picture of the structural formula of a phosphate molecule possessing ability to penetrate through soils flashed before my mind's eye, apparently without myself having had anything to do with it. My subconscious must have been working on the problem, while I wasn't.

That conversation between me and Dean Stewart took place on a day late in June, 1928.

Well, that was where, and when, and how I happened to discover a way to cause phosphatic fertilizer to penetrate downward into the deeper soil.

14

Case Study in International Defense Systems: A Systems Analysis for the Prevention of Korean Demilitarized Zone Inadvertent Overflight*

WILLIAM M. KINNEY

First Lieutenant William M. Kinney, United States Army, wrote this paper for a USC/MSSM course at Yongsan Army Garrison, Korea in the winter of 1977. Lt. Kinney, at the time, was a UH-1 Helicopter Aviator qualified to fly in the aviation buffer zone adjoining the southern boundary of the demilitarized zone (DMZ) between North and South Korea.

The problem of inadvertent overflights from the South to the North has vexed the United Nations Command leadership since the establishment of the Military Demarcation Line across the peninsula of Korea in 1953. Lt. Kinney's study was submitted to the United Nations Command and Eighth U.S. Army leadership and received official endorsement for further feasibility study.

The case study illustrates:

1. An original conceptualization for the solution to a critical, long-lasting, problem in international civil-military affairs.
2. A systems analysis on level 6 of Table 4.2 (i.e. "Prescribe a system improvement, design, or redesign to resolve a specific problem—mostly normative research; assumes in-house possession of behavioral knowledge").

Acknowledgements: Sincere appreciation is given to the Staff of the United Nations Armistice Commission in Yongsan, South Korea for their help in gathering critical data utilized in this analysis. Special thanks is given to Mr. Jimmy Lee, Chief, Historical Branch for his personal efforts in securing the historical data that forms the basis for this study; his advice and observations were extremely helpful in formulating the remarks found in the political analysis section.

Special recognition is gratefully given to the men of the Operations Section, 128th Aviation Company (Assault Helicopter), Uijohgbu, South Korea. Their support and help in the gathering of information was a critical determinant in the quality of research accomplished.

3. The perception of technological and economic feasibility influencing political feasibility of responsible leadership to seriously consider the study.
4. The principle of system quality leverage (see Table 7.2) through a small analytical input having a potential for significant systems improvement.

R. M. KRONE

INTRODUCTION

This paper is a systems analysis of helicopter operations in or near the Demilitarized Zone in South Korea. Its specific intent is to devise a viable system for the prevention of air violations of the Demilitarized Zone during military aviation operations conducted in the Republic of Korea. The methodology of study utilizes historical data to document the continuing occurrence of air violations throughout the twenty-four year Armistice. The present system of visual controls is analyzed and found to be inadequate for modern military aviation tactics and training doctrine. A values analysis and political assessment of an air violation is conducted using a recent example as a poignant reference for discussion. Several options are elucidated but disregarded for violating either technological, political, economic, or military feasibility criteria. A recommended addition to the existing control structure is proposed by advocating the modification of a present air-to-air proximity warning device to a ground-to-aid control measure. A feasibility analysis is conducted and considers technological compatibility, terrain analysis, economic feasibility, and political connotations. The proposed addition appears to meet the feasibility criteria and is highly recommended.

This analysis was motivated by the inherent hazard of aviation operations in Korea as experienced by the author and documented in the history of Armistice Violations. The criticality of preventing overflights of the hostile border between North and South Korea was again highlighted in July 1977 when a CH-47 "Chinook" helicopter strayed across the border and was shot down. As with all previous inadvertent DMZ overflights from south to north there was a high cost in human, material, political, and military terms.

THE HISTORY OF AERIAL ARMISTICE VIOLATIONS

Since the implementation of the Military Demarcation Line (MDL) across the peninsula of Korea in 1953 there have been numerous violations or charges of violations against both sides. Table 14.1 provides the data by year separated into violations charged by one side and those admitted by the other side. An admission by one side in response to the other side's charges is the basis for an official violation. Violations charged by one side but not admitted by the other remain unofficial and have impacts for international politics and propaganda but do not become agreed agenda items for negotiations at the Joint Security Area (JSA) at Panmunjom where all North-South dialogue occurs under terms of the 1953 Armistice Agreement.

Table 14.1 reveals that during the entire 24 year history of the MDL, the North Koreans have never admitted to an aerial violation of the DMZ. On the other hand, under the section involving charges against the United Nations Command admissions are frequent. There have only been 5 years in which air violations have not been admitted by the United Nations Command through 1977. Of the total air, ground, and sea violations admitted to by the United Nations Command, 80% have

Table 14.1 *Armistice Aerial Violations of the DMZ**

| Year | UNC[†] against the KPA/CPV[‡] | | KPA/CPV against UNC | |
	Charged by UNC	Admitted by KPA/CPV	Charged by KPA/CPV	Admitted by UNC
1953	28	0	135	13
1954	20	0	261	13
1955	12	0	100	4
1956	2	0	19	2
1957	9	0	55	0
1958	7	0	44	7
1959	1	0	13	0
1960	0	0	19	9
1961	5	0	13	6
1962	0	0	15	2
1963	0	0	14	6
1964	0	0	14	5
1965	2	0	6	3
1966	0	0	1	1
1967	1	0	17	1
1968	1	0	18	2
1969	1	0	10	1
1970	1	0	16	1
1971	0	0	20	1
1972	0	0	8	1
1973	0	0	11	0
1974	0	0	38	0
1974	0	0	38	0
1975	15	0	46	1
1976	1	0	84	0
1977	0	0	34	1

*Source: United Nations Armistice Commission, Youngsan Army Garrison, Seoul, Korea. Data current as of July 1977.
[†] UNC = United Nations Command
[‡] KPA/CPV = Korean Peoples Army/Chinese Peoples Volunteers

been air violations. This clearly substantiates the need for intensive analysis and productive recommendations to end this continuing source of political embarrassment and to design an effective control system to safeguard the lives of pilots, crewchiefs, and passengers and prevent the loss of valuable aircraft.*

Figure 14.1 depicts the geographic location of air violation occurrences by assigning the violations to one of seven established Buffer Zones extending from west to east across the Korean Peninsula at approximately the 38th parallel. A radar facility monitoring Buffer Zone flights in Area II has been a factor in the limitation to four violations in that zone. Due to the low altitude flights conducted by helicopters, however, the effectiveness of this radar is largely confined to the open spaces of terrain within the line-of-sight capability of the radar. The correspondingly low number of overflights in the eastern areas (VI and VII) is more a function of the lower number of flights in these areas rather than improved control techniques or greater visibility of the DMZ (referred to by pilots as "the fence"). Conversely, there appears to be a higher proportion of air violations in the area of the Han River Estuary in Buffer Zone Area I. It is reasonable to assume that this is a function of the lack of terrain references since there is no visible "fence" to delineate the southern boundary of the Demilitarized Zone. All of the air violations occurred during the period of daylight and all but one were made by military aircraft. Flights into the Buffer Zone between the times of local sunset and local sunrise are prohibited except for emergency medical evacuation.

PRESENT CONTROL SYSTEM

In this section the discussion will focus on only the relevant aspects of the control procedures utilized in monitoring Buffer Zone Flights. The major reference for this discussion will be United Nations Command/U.S. Forces Korea/Eighth Army (UNC/USFK/EA) Regulation 95-3 (dated 1 July 1975). This regulation establishes flight restrictions, procedures, and aviation training requirements to provide the Commander, Eighth United States Army with control of all Eighth U.S. Army, Republic of Korea Army (ROKA), and 1st ROK Marine Division flights within or through the Korea Tactical Zone and Buffer Zone.

At this time it will be helpful to explain the terminology to be used in this discussion as it pertains to the Buffer Zone. The reader is encouraged to refer to Figure 14.1 for a pictorial display of the Buffer Zone. The Military Demarcation Line (MDL) is a line across the Korean Peninsula separating the area under military control of the United Nations Command Forces from that controlled by the North Korean Peoples Army. The Demilitarized Zone (DMZ) is an area 4000 m in width extending approximately 150 miles across Korea from the Han River Estuary

*For a short description of the procedures, controls, and training for pilots who fly in the Buffer Zone see, Captain Erick L. Mitchell, "A Helicopter Flight in the Korean Buffer Zone (BZ)," *United States Army Aviation Digest* (March 1979), Vol 25, No. 3, pp. 38–39.
†UNC = United Nations Command.
‡KPA/CPV = Korean Peoples Army/Chinese People's Volunteers.

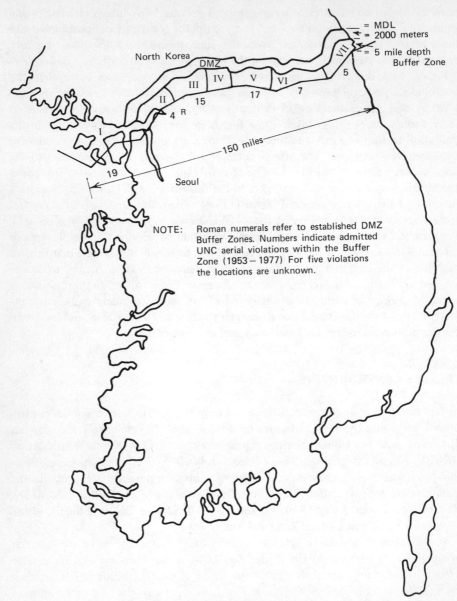

Figure 14.1 South Korean DMZ buffer zones.

(on the western coast) to the Eastern Sea. The general trace of the southern boundary of the DMZ is approximately 2,000 m south of the MDL. The northern boundary is 2,000 m north of the MDL.

The code word "Waterloo" is used by air traffic control and air defense communications facilities to clear the Buffer Zone airspace to prevent a possible DMZ overflight, to declare an air defense alert, or for any other emergency which would require clearing any particular area of the Buffer Zone. The term "flight following" is used to indicate the requirement to monitor the flight progress of aircraft and is normally associated with air traffic control. "Radar monitoring" is a service provided by a radar facility to an aircraft when the primary aircraft navigation is being performed by the pilot. This service involves the continuous tracking and reporting of an aircraft in flight and advises the pilot of potential flight hazards, provides navigational assistance as required, and relays information.

All aircraft are required to maintain two-way radio communication while in the Buffer Zone as well as having operable radio navigation equipment. There is, however, a dearth of radio navigational aids and those few now available are subject to meaconing (jamming) practices from the North. The other control measures incorporated into the present system are a string of "T" markers 2000 meters from the southern boundary of the DMZ. These markers are formed by rocks, painted yellow, approximately 2000 m apart; their dimensions are approximately 8 by 10 feet. They are placed so that the horizontal portion of the "T" is at the northern end and are placed so as to warn pilots of their proximity to the DMZ. Notwithstanding the intention of their design, they are difficult, if not impossible, to see unless one flies directly over them.

The other set of control measures are a series of 33 checkpoints across the Korean peninsula. These checkpoints are just specific terrain features (road intersections, hill tops, river bends, etc.); they are used to describe a route of flight required when the aircraft is north of the Buffer Zone line. They are mandatory reporting points to report the position of the aircraft to the flight following agencies. Finally, along the fence, there are guard posts continuously manned and these individuals are supposed to fire red pyrotechnics in the event of an imminent DMZ overflight. These posts are spread out intermittently along the southern boundary of the DMZ and they rely heavily upon the practice of waving a small red flag to warn approaching aircraft. The fence, it should be noted, is not a reliable navigational aid since in some areas it is located north of the original southern boundary of the DMZ.

Not all aviators assigned in Korea become Buffer Zone qualified, nor should they be. Only a select few based on experience, judgement, and maturity are selected from each unit depending upon the mission requirements of that unit. Further, U. S. Army Aviators are only trained in Buffer Zone Areas I–IV due to the role of the U. S. Forces in the event of renewed hostilities. U. S. Army helicopters can, of course, fly in Areas V–VII, but they must first stop and pick up a ROK aviator who will then act as a navigator.

The training required of U. S. Army Aviators by the regulation is quite rigorous and detailed. At the end of such training they are required to take a checkride and demonstrate their proficiency at (1) flying the imaginary Buffer Zone line by reference to terrain features, (2) fly the line connecting the checkpoints by reference to terrain features, and (3) fly along the fence. It would seem that such training results in highly qualified individuals and this author has no hard data to dispute that contention.

There are some other significant variables to be considered. First, although the weather criteria for flights into the Buffer Zone can be no less than cloud ceilings

500 feet above ground level and a visibility of 2 miles, there is a small margin for human error in such weather conditions. The regulations and training pertaining to Buffer Zone flights place a great deal of emphasis on "meticulous pilotage" and map reading ability, not entirely a foolproof method given the human factor. In addition to the regulations, various Commanders and current aviation training doctrine encourage flights at low level to make the most effective use of the terrain to increase aircraft survivability and mission accomplishment.

There is a stress on low-level flight in helicopters according to the current training doctrines. Hence, there is a need to train in the Buffer Zone at the lowest altitude consistent with safety of flight and navigational considerations. Thus, in the event of renewed hostilities and the sophisticated air defense environment that will accompany it, mission essential flights could be made successfully. A second reason for low-level flying in the DMZ is to deny to the North Korean military valuable intelligence about aircraft and key personnel movements within the Buffer Zone by reducing the ability of the North to monitor, by radar, aircraft location and gain other intelligence by radio monitoring. If this is the case, then our single radar facility which monitors Area I–III will be similarly hampered and in effect would function only as a radio flight following facility. Similarly, the visual aids discussed earlier (the "T" markers and red flag-waving) are greatly reduced in effectiveness especially for aircraft traveling at speeds in excess of 90 knots; the cruise speed for the UH-1 helicopter.

Based on data obtained for a 12-month period ending October 1977 dealing with Buffer Line clearances and radio transmissions, it is my estimate that there are approximately 1000 rotary wing and 250 fixed wing flights into the Buffer Zone every month. It is my opinion that the current system of controls is primitive compared to the high density of traffic transiting this highly volatile area. The control measures certainly do not measure up to modern tactical doctrine, nor do they provide a technological match for aviation assets in Korea.

VALUES ANALYSIS

Values are things or principles preferred (see Chapters 5 and 6). The protection of human life is high on every list of preferences. The air violation of the CH-47 in July 1977 resulted in the loss of three crewmembers and the aircraft. The dollar value of a CH-47 is $2,956,896.

A message was disseminated to all aviation units in Korea from the highest levels of the military command structure reemphasizing the criticality of professionalism while performing aviation duties in the explosive environment of Korea. From top management came their values analysis; "The loss of the CH-47 and three crewmembers was too high a price to pay . . ."

It is incumbent upon all commanders to safeguard the lives and property entrusted to them by virtue of their rank and position. In regards to the current control procedures utilized in the Buffer Zone, is it enough to just stress "meticulous pilotage"? To be sure, Army aviators are all familiar with map reading and some are quite accomplished at it. But is everything that could conceivably be done in the light of low-level tactics and marginal weather flying conditions being done? I think not.

The purpose of all training is to provide increased skill to the person receiving it. Although I believe the individual unit training to be adequate, we still leave a gap in our control processes: the human factor. Are the conditions proper so that one can safely assume that a high degree of infallibility will result when aircraft fly in the Buffer Zone? Our training and navigational aids to assist the pilot in conducting error-free navigation do not meet the standard of "prevention of inadvertent DMZ overflights" (see Chapter 7 for the Krone discussion of standards). The evidence is summarized in Table 14.1.

POLITICAL CONSIDERATIONS

In examining the political ramifications of DMZ overflights throughout the history of the 1953 Armistice one can easily grasp the idea that such airspace violations have resulted in the loss of much "political face." In the political forum at Panmunjom and in the international press, the United States and the Republic of Korea have been ridiculed, made the object of calumny and vociferous propaganda, and have been looked upon as aggressors from the south.

In the 1977 incident, the reverberations went to the highest political and military levels of concern. For a few days in July, Korea stood tensed—on the brink of political-military escalation. Skillful diplomacy averted a renewal of hostilities. During the alert, however, large numbers of men and material were mobilized at a significant cost. If such is the political visibility of an air violation, and if the historical data is accurate, then it is imperative that greater work be done to prevent history from repeating itself.

This becomes even more crucial with the potential American military withdrawal as American troop reduction will lower the level of deterrence. There would be less margin remaining for an air violation to avoid rapid escalation into renewed warfare. Such combined political military extrarational considerations must be included in the systems analysis of the current control system to prevent DMZ air violations (see the Krone discussion of extrarational variables for systems analysis, Chapter 5).

ALTERNATIVES

There are several options that could be considered by a decisionmaker, including the option of doing nothing and allowing the high-risk status quo to remain in effect; an option which this analysis does not recommend.

The second option is that of prohibiting all flight in the Buffer Zone. This option suffers a lack of feasibility due to the criticality of helicopter operations in the Republic of Korea. Helicopters provide maximum flexibility to key personnel in the performance of their duties as well as providing a superior vehicle for the quick and efficient allocation of military resources to critical locations. Supply and resupply activities to remote locations would greatly suffer. Given the existing transportation structure in the Republic of Korea, the feasibility of removing helicopters from the Buffer Zone is small.

Furthermore, an assumption behind banning all Buffer Zone flights would be that it is Buffer Zone qualified pilots who pose the greatest risk of overflights. I do not believe that such is the case. Although I have no hard evidence to support it. I believe that the greatest threat to inadvertent overflights comes from non-Buffer Zone qualified pilots transiting the Tactical Zone to the south of the Buffer Zone. In either case, a control system for both should be devised.

A third option for the decisionmaker would be to prohibit all low-level flight. Such a measure would negate the need for low-level flight training but would be inconsistent with existing army aviation doctrine designed for decreasing higher flight altitudes; also it would provide increased intelligence to the North Korean military. This option if implemented would lend itself to establishment of radar contact at all times. At least two more expensive radar facilities would be needed along the Buffer Zone, also the requirement for positive radar contact has several inherent problems. There could be no flights into the Buffer Zone during periods of maintenance or radar downtime. Secondly, the weather criteria for Buffer Zone operations would have to be raised significantly due to the terrain clearance requirements, especially in the eastern areas of the peninsula. Thirdly, it is common Air Traffic Controller practice to maintain a minimum of five miles separation between aircraft under their control and this requirement would be impractical in the five mile deep Buffer Zone. All in all, this option should be rejected due to its inherent inflexibility and exorbitant cost. The fourth option—providing a cockpit signal for all flights operating in the Buffer Zone—is my recommended solution.

RECOMMENDED SOLUTION

The first characteristic I looked for in a feasible system to prevent border overflights was a system that would augment, enhance, and act as a failsafe device for the pilot whose primary means of navigation would continue of necessity, to be "meticulous pilotage navigation." Such a system would further have to be as flexible as the aircraft in its low-level routes of flight. That is to say, the overall system should allow low-altitude flight operations in valleys as well as over the crests of ridges and mountains. Third, its implementation and operation must be simple to be cost-effective. A further critical constraint was that due to time and monetary restraints any technological option chosen had to be currently available for immediate procurement. A systems component whose design was to go beyond the state of the art was discarded. I was interested in a system whose control measures provided pilots with immediate cockpit indications so that corrective action could be initiated promptly.

My recommended solution is a series of marker beacons across the peninsula.

Marker beacons serve to identify a particular location in space along an airway or on the approach to an instrument runway. This is done by means of a 75-MHz transmitter that transmits a directional signal to be received by aircraft flying overhead. These markers are generally used in conjunction with enroute navigational aids and the Instrument Landing Systems as point designators. At 100 feet above the beacon an elliptical beam is produced that is 2 to 3 miles in length and approximately 1000 feet wide. At 500 feet above the beacon the beam is 8 miles in length and 1 mile wide. At 1000 feet above the beacon, the beam is 12 miles in length and 4 miles wide. The marker beacon unit is powered by 110/220-V alternating current that could be made available in the southern Buffer Zone. My rough estimate is that between 30 and 40 units would be required for installation across the Buffer Zone line of approximately 150 miles.

With the marker beacon receiver in all aircraft flying in South Korea a visual and an oral signal could be generated in the cockpit to alert the pilot that the aircraft was entering the Buffer Zone. If this were an inadvertent overflight even the most confused pilot could turn the aircraft to a heading of south and avoid flying further north across the Demarcation Line into North Korea. This would be possible even if the aircraft compass system were inoperable as all aircraft have reliable standby compasses that are independent of all power sources. If the flight was one planned for the Buffer Zone the oral signal could be turned off and the visual one (a cockpit red light) could be dimmed or merely used as an additional navigational check for the aircraft entering and leaving the Buffer Zone.

The technology to receive the marker beacon signal is already installed in the UH-1 and CH-47 helicopters and all components of the system are current state of the art and available in the inventory. If military or political decisions called for it these units could be easily removed after installation. If pilot proficiency should ever be lower than it is now this system would be an important hedge against the higher probability of inadvertent overflights. Although an in depth economic feasibility study is beyond the scope of this paper it is estimated that the entire system—ground and airborne equipment—could be installed and operating at a cost of less than $500,000.

RECOMMENDATION

I recommend that the Eighth United States Army conduct an in-depth feasibility study into the possibility of installing a Marker Beacon electronic fence (or preferable alternative should one exist) along the southern portion of the Buffer Zone to the Military Demarcation Line between North and South Korea to provide an improved system for the prevention of inadvertent overflights from south to north.

15

Case Study in Automation: Automatic Packing in a Pharmaceutical Firm

CAROLINE CHUNG, HEINZ KESTERMANN, AND ARTHUR SOMMERS

This study was accomplished for my August–September 1977 "Systems Analysis" class at the USC Study Center in Taipei, Republic of China. Dr. Heinz Kestermann was the General Manager of the pharmaceutical company studied. Captain Arthur Sommers, U.S. Army, was stationed in Taiwan, and Miss Caroline Chung was a citizen of the Republic of China. All three were USC/MSSM graduate students. The study illustrates:

1. Behavioral, values, and normative research.
2. Analysis of economic, technical, and political feasibility.
3. Computer model applications for comparative cost-benefit projections on which corporate management can base its planning.

This study was submitted to top management of the pharmaceutical firm parent corporation and was considered a valuable input for corporate planning.

R. M. KRONE

INTRODUCTION

The world we live in today is a rapidly advancing technological complex of interactions between man and machine. As a country gains economic prominence, the people of that country desire greater rewards. It is based on this premise that oftentimes human labor must be replaced by machines for several purposes. First the manpower market becomes priced so high that a company can no longer compete on the local or international market. Second with the price of real estate increasing, the expansion of an organization in terms of real estate is often not feasible and the organization must automate to optimize the utilization of existing space. Third, as the organization grows it is imperative that more innovative techniques of forecasting and budgeting be adapted which often cannot be accomplished based totally on qualitative variables to the exclusion of quantitative ones (see Chapter 8).

The intent of this paper is to evaluate the benefits and costs of replacing a semiautomatic process (manpower) with a fully automated process (machine). It is our intent to show the relationship of manpower costs versus machine costs based on economic, political, and technological feasibility as related to a particular firm operating in Taiwan.

The German Remedies Taiwan Ltd. (GRTL), a pharmaceutical manufacturer, is a joint venture by two German pharmaceutical companies. The approval for the foreign investment by the two companies was given in 1967 with actual production not starting until 1972. In 1974, a Swiss company was accepted into the technical corporation agreement. As of 1976, approximately 70 percent of output was sold to local markets in Taiwan and the remaining 30 percent was exported to affiliated companies in Hong Kong, Malaysia, and Singapore.

The company makes a wide variety of products, which are packed in accordance with the desires of each of three sponsor companies based on market analysis. The production output of the company is in the form of tablets, dragees, capsules, syrups, ointments, and injectables. The majority of the output (85%) are in the form of tablets, dragees, and capsules. The packing of the presentations encompasses a wide variety of containers; the smallest packs are two pieces in one carton to the largest of 1000 tablets in bottles. Quality and presentations are in line with the official recognized regulations of the World Health Organization (WHO) and the parent companies. The packing of the product must not only ensure protection from damage but must also ensure that the active ingredients of the product remains stable for the amount of time specified, usually 5 years.

Our study is limited to the packing of dry presentations, that means, tablets, dragees, and capsules, which represent about 85% of the total output. In 1976, GRTL manufactured 27 products of this type to be packed in 89 different packings. Sales of the smallest product are about 30,000 pieces per year, the biggest about 14 million per year. There are three types of packing; bottle pack, stripping in aluminum foil, and blistering in PVC/aluminum foil. In 1976, about 20 operators packed 63 million pieces in 38,300 working hours. Total cost of the packing department during this period was about 3.7 million New Taiwan Dollars (NT$) (about 97,000 US$) out of which about 30% are personnel costs. For cost calculation purpose, the following procedure is adopted. All operations expenses are allocated to seven productive cost centers. The total cost of one center is divided by the productive manhours, leading to the cost per hour. The definition of the term productive manhour is the responsibility of top management. As the company works on a manual and semiautomatic basis, the manhours are relevant factors in cost calculation. If there would be an automatic line, our cost of this equipment would be the basis for the calculation. The essential part of the production is the delivery of oral contraceptives to the Taiwan Family Planning Association and deliveries have to be made normally on short notice. Because of this short notice, an automatic blistering machine had to be purchased in 1976.

The capacity of the machine is not fully utilized. However, the machine enables GRTL to meet the delivery schedule of the Family Planning Association.

BEHAVIORAL RESEARCH

In the past Taiwan has always been considered as a country with ample resources of cheap labor and having the ability to adopt the skill required for the mass manufacture of technical products (e.g., electronics, spare parts, textiles, shoes). Accordingly, the country attracted foreign investors making profitable use of resources. Recently, however, official Republic of China announcements have been made about apparent labor shortages. In addition to the shortage, industrialization has produced a rise of living standards, which has caused a considerable increase of wages and salaries.

When analyzing manpower direct charges and sales have to be taken into consideration plus the indirect costs of management of the company and compliance with government regulations, covering vocational training, welfare funds, retirement schemes, and other benefits. Working hours are also being shortened. There is ample evidence to show that Taiwan is rapidly losing its low labor cost business advantage compared with some other Asian countries.

The assurance of superior quality product is an important factor for the pharmaceutical industry. Although employee training is relatively easy human errors cannot be excluded. With the large variety of products packed in GRTL, considerable supervisory effort is required for quality assurance in the packing department where several products are packed simultaneously. This supervision increases costs.

Export is an essential part of GRTL's business. Accurate forecasting and budgeting is of utmost importance as quotations must be made well in advance of deliveries. This capability is negatively influenced by the great percentage of personnel cost in the packing department as salaries increase due to inflation and unpredicted labor shortages occur. Market fluctuations must also be taken into consideration.

Another export variable is that the work load is not equally distributed throughout the year. There are peak demand periods, which means that the number of packing girls should vary throughout the year according to the demand. Although periodic layoff and rehiring is possible it can have a negative impact on the factory environment and is counter-cultural for Taiwan business enterprises which traditionally take a family attitude toward employees.

Pilferage has not existed in the last five years and is predicted to be a variable.

The contamination of the products in the packing department is of utmost concern. With the pharmaceutical presentation being in an unprotected state at arrival to the department, constant checks must be performed to insure personnel and facility sanitation. The World Health Organization (WHO) has recently promulgated new regulations defining more stringent standards for the prevention of contamination.

THE EXISTING SYSTEM

The existing system is comprised of various semiautomatic and manual processes for each type of packing. The packing room is approximately 75 by 25 ft and is at capacity with current equipment and personnel; to achieve an increased output would either require expansion of the facility or going to a two shift operation. Currently the packing department operates from 0800 to 1630 hours Monday through Friday, unless special orders require additional hours. The department, like the rest of the plant, is a closely controlled environment with very strict sanitary regulations enforced. The three basic dry presentation packs are the bottles, strips, and blisters. Following is a description of each.

The bottling process is sensitive to lot sizes with the smallest being 10 capsules per bottle and the largest 1000 tablets per bottle. The bottles are sterilized and then put on a labeling machine where a label is automatically glued to the outer surface. The bottles are then sent to the packing lines where filling takes place. For the smaller lot bottles (10, 25, 50, up to 100) the filling is accomplished by an automatic counting machine which releases the desired amount of pills as the operator presses the bottle to the filling nozzle. For the medium lot bottles (100 to 500), the filling is accomplished by a manual operation, where a matrix is used. This matrix is designed so that as a hopper of pills crosses it the desired amount of pills will fall into holes cut to the size of the pill. The operator then pulls the hopper away and visually checks to insure all holes are filled. Another operator will then release the pills into a bottle. For large lot bottles (500-1000) the bottles are filled by weight. The operator holds the bottle to an automatic dispensing machine until the desired weight is achieved. It should be noted that often this process gives a few more pills than the desired amount, but management feels this is more economical than counting the desired amount.

The bottles are then sent down a conveyor belt where other workers put in a small foam rubber buffer to prevent movement of the pills during shipment and a cap is affixed. The bottle continues down the conveyor and is put into a box with a small instruction sheet. The box is then put into a carton and sealed for shipment.

STRIPS

The stripping machine is a semiautomatic process, which seals individual pills between two layers of aluminum foil. This process is accomplished by placing pills in a hopper that has a rotating bottom with holes cut to the size of the pill. As the bottom rotates it picks up the desired amount of pills and releases these pills into nozzles that place the pill between the aluminum foil as it also rotates. Once the pill is between the foil it is heat sealed. The machine cuts the strips into desired amounts and the packet falls into a box. All packets are then manually checked to insure each packet contains a pill and excess foil is manually cut away. The strips are then put on a conveyor belt where workers place the desired number of strips into a box with an instruction sheet. The box is placed on a conveyor belt and is weighed prior to going into a shipping carton.

BLISTER

The blister machine was acquired to meet the rising demand for oral contraceptives. The machine in its current configuration requires little manual intervention. The pills are placed in a hopper and gravity fed into molds that accept the desired amount (21 or 28 pills). The PVC material is rotated through a heating device and is then pulled across a suction machine that forms pockets in the material into which the pills will fall. With the pills in place the top aluminum foil cover is affixed by a heat process. The packet is then cut to the desired shape and dispensed into a box. The box is moved to a conveyor line where operators visually inspect to insure all packets are filled and then place the packet inside an aluminum bag, which has been precut. These bags are then heat sealed and placed in a box and sent to the next conveyor line. On this line the packets are counted and placed in a box with an instruction sheet. The box moves to the end of the conveyor belt where it is weighed and put in shipping cartons.

Although the above description of the existing system has been simplified it gives an idea of the current cumbersome methods being employed. It also shows that a direct relationship now exists between manpower and output. This means that as output requirements increase so do manpower requirements.

THE PROBLEM

Our behavioral analysis reveals a need to look into the possibility of replacing the existing human work force by a fully automatic packing device. Automation may lead to decreased flexibility, which is an important capability for GRTL due to its diversified products. To fully automate the existing operation the following equipment would be required in addition to the already existing blister machine: cartoning machine, several control devices, photocells, transfer equipment, and spare parts. The output of the additional equipment must exceed the output of the existing system. By operating such an automated packing line seventeen workers could be eliminated leaving three operators for the entire operation. The hypothesis motivating this research is that automation of the packing department will result in (1) improved product packaging quality, (2) decreased labor cost, and (3) increased capability for forecasting and budgeting requirements of GRTL.

THE PROPOSED AUTOMATED SYSTEM

The new system would be a fully automated operating line with the following components: control box; forming station; product feed; filling section; memory bank; photocell; sealing station; perforation station; gripper transport; unwinding of cover foil; waste coiling unit; punching station; transfer system and cartoning machine. The existing blister machine could be utilized if the new system input were aligned with the blister machine output. The photocell checks all the pockets of the blister to insure they are filled with unbroken pieces. The memory bank counts the output and gives impulses for ejecting defective packets identified by the photocell. A transfer system would be needed to either convey the packet to the cartoning machine or to the aluminum bag machine and then on to the cartoning machine. The aluminum bag machine would place the individual packets into an aluminum bag and seal the bag.

ADVANTAGES AND DISADVANTAGES
OF THE PROPOSED SYSTEM

The system could be operated by three employees including the packing in cartons for shipment. Product quality would be improved by minimizing the risk of inadvertent product mix. Sufficient control devices would be installed on the proposed system thereby sorting out defective packs, which could be overlooked by visual checking. Peak demands could easily be met by running the proposed machine more than 8 hours per day without additional cost for overtime. Personnel problems would be reduced due to smaller work force. The forecasting and budgeting would be more exact as expenses involved in running a machine are easier to be forecasted than wages and salaries and other expenses related to personnel. The same parent company that controls GRDL is also in control of a company in Japan that has adopted the total automated blister packing line with considerable success.

As for the disadvantages there would be only one kind of packing for tablets, dragees, and capsules, the consequences of which would have to be investigated from the marketing point of view. The PVC foil for blistering has to be imported as a suitable quality is not available locally. At present GRTL is able to import the required foil quantities, however, for the investment under discussion, GRTL needs the guarantee that this foil will be available in the next 5 to 8 years. Should the government of the Republic of China impose import constraints the innovation under investigation would become useless. Since a governmental guarantee for a long-term import license is not now politically feasible local foil manufacturers should be approached to determine the likelihood of suitable local supplies. All other packing materials have to be purchased locally including the outer boxes. Based on the experience of another pharmaceutical manufacturer in Taiwan the quality of locally procured outer boxes is below that necessary for automatic cartoning. Imports would involve additional freight costs and import duty.

During periods of power outages such as those caused by typhoons, the packing department of GRTL would have no output. This risk seems acceptable based on past experience with power failures.

The new system would require about 9 months for installation. That time could be used to reduce the existing staff by means of normal personnel attrition. In addition, some packing line employees could be transferred to other departments. Termination of the remaining personnel should not be a major problem considering Taiwan labor needs. The resale possibilities for the existing semiautomatic equipment are minimal.

TECHNOLOGICAL FEASIBILITY

The automation of the packing department is technologically feasible. Alternative equipment proposals were evaluated. The most suitable equipment costs 4.8 million NT$ in 1977 (thirty five NT$ equalled one U.S. dollar) with an installation time of 9 months. Sufficient space is available in the existing packing department to house the additional equipment. The required power is available. Training of both operator and maintenance personnel can be accomplished by the manufacturer during a time period to coincide with installation. One area that will warrant further study is a problem with the boxing material and the cartoning machine. The cartoning machine is designed to accept a flat box and it will bend the box in a manner to allow the packets and instruction sheet to be inserted and then automatically close the box. Information from other pharmaceutical firms indicate that the boxes procured locally will not work in this automatic machine due to inferior quality material. The boxes are to be prefolded, but are often not, therefore the machine will fold it in the wrong place, which will not allow the packet to be inserted. This problem may necessitate the firm to import more costly boxing material from outside Taiwan.

ECONOMIC FEASIBILITY

The cost-benefit analysis of the proposed packing system is based on comparisons with packing cost per unit in current semiautomatic operations. The sum of manhour cost and material cost is the packing cost of the existing system. In the proposed system, packing cost is the sum of machine-hour cost, material cost, and system change time cost. "Manhour cost" or "Machine-hour cost" becomes a productive hour cost standard or the current system and proposed system, respectively (see the discussion of standards in Chapter 7).

Past data for personnel, energy, depreciation, interest, operating material, and repairs for the years 1972–1976 were analyzed. These elements are the secondary criterion assumed to be positively correlated to the primary criterion of "productive hour cost." Certain costs, such as quality control and warehouse costs, were not included in this analysis. For projection purposes through the 1980 time frame a trend analysis was done based on the 1972–1976 data. In the trend analysis we decided to use the least squares method. Our first attempt utilized the exponential trend analysis, but after considerable evaluation we felt that this projection did not align with the company's goals. The least squares method uses a linear model that does align with the company's growth projection for the next 5 years. The proposed system cost estimates were developed for personnel, energy, depreciation, interest, operating material, and repair; (based on 1976 costing guides) then utilizing the same percentage increase as the existing system, we projected this proposed system to the year 1980. Table 15.1 shows the base costs used (1976) and the annual projections through 1980. Costs in Table 15.1 are projected based on the medium term sales forecast set by the companies goals which in turn, was based on a total market analysis, the companies funds to be allocated, economic fluctuations projected, and the introduction of new product lines to be phased in during certain time periods. Table 15.2 contains calculations of the cost per unit for the existing and proposed system.

Tables 15.1 and 15.2 show the break even point for the existing and proposed system. Equal total costs or equal costs/unit is where the use of one system has no benefits over the other system. The concern then turns to the future and as noted on the projection, the proposed system will be much cheaper to operate. From our analysis the optimum conversion date to the proposed system would have been in the latter part of 1976.

Now that our productive hour cost has been calculated and we have derived our per unit cost for both the proposed and existing system, (Table 15.2) we

Table 15.1 *Cost Comparison Existing/Proposed System*

Item	Cost per Year, NT$				
	1976	1977	1978	1979	1980
		Existing			
Personnel	1,200,000	1,358,400	1,572,000	1,785,792	1,998,301
Energy	300,000	393,000	472,650	552,527	632,091
Depreciation	400,000	469,600	540,800	612,185	683,199
Interest	400,000	469,600	540,800	612,185	683,199
Operating material	200,000	228,000	255,200	282,251	309,347
Repair	200,000	219,400	250,400	281,449	312,409
Total cost (NT$)	2,700,000	3,138,000	3,631,850	4,126,389	4,618,546
		Proposed			
Personnel	234,000	264,888	306,540	348,188	389,622
Energy	600,000	786,000	945,300	1,105,055	1,264,183
Depreciation	880,000	792,000	712,800	641,520	577,368
Interest	880,000	792,000	712,800	641,520	577,368
Operating material	100,000	114,000	127,600	141,125	154,673
Repair	200,000	248,000	296,000	392,000	488,000
Total cost (NT$)	2,894,000	3,001,888	3,061,040	3,269,408	3,451,214

Table 15.2 *Critical Data*

Item	Cost per Year, NT$				
	1976	1977	1978	1979	1980
	Existing system				
Total cost (NT$)	2,700,000	3,138,000	3,631,850	4,126,389	4,618,546
Production hours required	38,300	37,687	42,896	46,327	50,940
Units of output	4,200,000*	4,704,000	5,080,000	5,590,000	6,150,000
Cost/hour (manhour) (NT$)	70.5	83.27	84.7	89.07	90.63
Cost/unit (NT$)	0.64	0.67	0.72	0.74	0.75
	Proposed system				
Total cost (NT$)	2,894,000	3,001,888	3,061,040	3,269,408	3,451,214
Production hours required	1,360	1,517	1,639	1,803	1,983
Units of output	4,200,000	4,704,000	5,080,000	5,590,000	6,150,000
Cost/hour (machine) (NT$)	2,127	1,979	1,840	1,813	1,740
Cost/unit (NT$)	0.63	0.59	0.55	0.54	0.52

*Percent of increase based on stated company goals.

can analyze the cost of packing one unit. This is the cost from the time it appears in the department until it leaves in a finished form, in one of three states (bottle, strip, or blister). For this analysis the per unit cost (Table 15.2) was added to packing material cost estimated for the proposed system and calculated from past records for the existing system. The packing material cost includes the materials used as well as the waste that is inherent with the system. This waste includes that foil that is lost due to adjustment of the machines, and during quality control checks for inferior quality material as well as bottles that may be broken during the packing operation. A summary of packing costs for the existing and proposed system is presented in Table 15-3. Provided are the costs for three different lot sizes. The small quantity includes those products where less than 0.5 million pieces per year are packed. The medium quantity includes products where output per year is 0.5 to 1 million pieces. The large quantity includes products where output per year is in excess of 1 million pieces. Further, the tables are broken into packing types—blister, bottle, and strips. This breakout by lot size was done to account for the variation in prices of packing small lots versus the larger lots. The manpower and use of machines become more efficient when working on one product for a long time whereas the smaller lots require considerable change time between product lines.

Table 15.3 *Packing Cost Comparison—Existing/Proposed*

Item	Cost per item, NT$*				
	1976	1977	1978	1979	1980
Bottles					
Small	7.65/ 6.61	8.46/ 6.69	8.99/ 6.73	9.63/ 6.93	10.23/ 7.06
Medium	17.91/64.24	19.82/65.12	21.05/65.56	22.56/68.20	23.98/70.40
Large	23.51/65.95	25.89/66.92	27.59/67.88	29.62/70.78	31.59/73.19
Strips					
Small	4.50/ 1.56	4.94/ 1.57	5.27/ 1.59	5.66/ 1.63	6.04/ 1.66
Medium	1.99/ 1.46	2.20/ 1.48	2.23/ 1.49	2.49/ 1.55	2.64/ 1.60
Large	1.84/ 1.13	2.02/ 1.14	2.15/ 1.16	2.30/ 1.21	2.44/ 1.25
Blister					
Small	4.06/ 2.70	4.47/ 2.72	4.76/ 2.73	5.10/ 2.78	5.44/ 2.81
Medium	—	—	—	—	—
Large	2.81/ 2.07	3.10/ 2.09	3.30/ 2.11	3.53/ 2.17	3.76/ 2.22

Batch size description

Small Less than 0.5 million pieces per year packed
Medium 0.5 to 1.0 million pieces per year packed
Large Greater than 1.0 million pieces per year packed

*Figures are in New Taiwan Dollars (NT$).

Table 15.3 shows that packing under an automated system would be beneficial for both strips and blisters in all lot sizes. In addition it shows that bottles are much cheaper under the existing system than under the proposed system in the medium and large lot sizes.

CONCLUSION

This analysis reveals both technological and economic feasibility for the installation of a new automated packing line for German Remedies Taiwan Ltd. (GRTL). Political feasibility was not analyzed. Comparative cost studies for the existing and proposed systems with projection to 1980 (Tables 15.1 and 15.2) show that in 1980 the proposed system will save the organization 1,167,332 NT$ over the existing system. Table 15.3 shows the benefits of the proposed system for blister, strips, and small bottles but not for medium or large bottles.

The problem of local manufacturers supplying suitable packing materials could lead to considerable disruptions in an automatic packing line. It is felt that all packing materials with the exception of PVC foil could be suitably supplied with enough pressure on the manufacturer. The PVC foil will require the pharmaceutical firm to find a suitable manufacturer and formulate a long-term contract to insure stability in the operation.

AREAS FOR FURTHER RESEARCH

Prior to a final top management decision to automate the GRTL packing process the marketing risks of selling only one type of packing should be further investigated. Some bottle packing may have to be considered as well. Furthermore, an investigation should be made to determine if ampoule packing could be included in the automatic processing. To obtain a more detailed cost analysis different packing presentations would have to be analyzed according to the quantity required.

The trend analysis for the economic feasibility study did not look at worst case situations. An analysis should be accomplished looking at the best and worst cases for various alternatives to automating the packing department. This analysis should include a look at the existing system, a fully automated system, and various mixes in between. If the fully automated system has a major failure of the magnitude warranting complete replacement, can the organization afford the down time? On the other hand can the company continue to be at the mercy of the labor market? These are worst case examples, but conversely what if the machine operates without failure for 5 years or the labor market becomes so plentiful that labor costs actually go down? An analysis technique that could be used is the matrix of best and worst cases. Each alternative should be very well defined and then each component (personnel, energies, etc.) should have values assigned for the best and worst case that could be reasonably formulated from studying past, present, and future projections for various like markets. This analysis would provide the decisionmaker more alternatives from which to choose as well as defining the uncertainties that would accompany that alternative.

If we rely totally on an automatic packing line then contingency plans should be formulated to allow for manual packing operation during times when portions of the line are down. For instance, should the cartoning machine go down personnel from other portions of the plant should be available for manually cartoning the product.

Locally produced packing materials should be evaluated to insure availability of ample supplies in the quality and quantity necessary.

RECOMMENDATIONS

German Remedies Taiwan Ltd. (GRTL) management should propose to the parent company's top management an expanded feasibility study for a fully automated packing line. If this is approved then the following measures should be taken:

1. Evaluate and carry out studies that warrant further consideration as identified in this paper.
2. Keep the number of packing department employees to a minimum, allowing normal personnel attrition to decrease personnel prior to implementation of the new operation.
3. Dispatch qualified personnel, who will be the mechanics for the proposed system, to other parent company plants in the Far East for training to minimize the cost of formal training.

16

Case Study in Management Information Systems : Information Systems and Corporate Structure

LARRY J. CAMPBELL

Mr. Larry J. Campbell was a systems analyst with RCA and a student in the USC/MSSM program at Kwajalein Island, Trust Territory of the Pacific when he accomplished this analysis in March 1977.

The study illustrates:

1. A rationale and analysis for converting from a manual management information system to a computerized information system in a medium sized company.
2. Use of explicated models with corporate leadership as client.
3. Corporate restructuring based on the new computerized information system.
4. A 5-year implementation plan and cost estimates.

R. M. KRONE

INTRODUCTION

This paper presents an analysis of the conversion of a medium-sized manufacturing company from a manual management information system to a computerized information system (CIS) drawing on the concepts and techniques presented in Chapters 4, 6, 7, and 8. Normative research is conducted to determine what should be, and concepts and analysis are presented for evaluation by decisionmakers.

The analysis is motivated by the potential benefits to be gained by the corporation in terms of decisionmaking corporate growth, return on capital, and other benefits that would result from the conversion. The existing features of the manual system are compared with the automated approach to show where the benefits and/or disadvantages occur. The total corporate information structure is analyzed to determine the area for initial concentration of efforts. Detailed analysis of the manufacturing data system is presented. An overall plan for implementation of a CIS is presented with recommendations for changes in corporate organization. Cost estimates are computed for the first 5 years.

This paper analyzes the Company's* existing information system and the impact of moving toward a computerized version of this system. The Company is a medium-sized manufacturer with two manufacturing sites located in the Midwestern United States. Four regional warehouses are located on the East and West coasts. Each warehouse services a sales region and several sales offices located in major cities on both coasts. The corporate headquarters is located at one of the manufacturing sites and houses the central corporate structure. Yearly sales are $100 million and are growing at about 12% per year. The Company is expected to begin more emphasis on international sales in one of the East coast regions.

Our organization is a classical business hierarchy (Figure 16.1) with major divisions grouped under vice presidents reporting to the company president. He in turn reports to the Board of Directors. The Company stock is traded over the counter and is controlled by the members of the Board, Vice Presidents, and employees. A small percentage of stock is owned by the public.

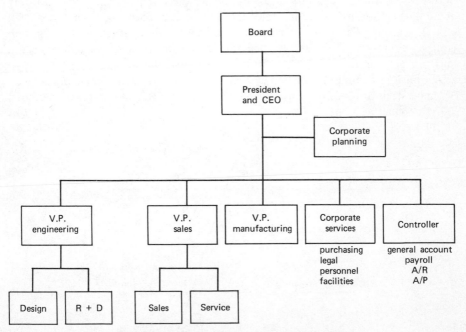

Figure 16.1 The existing company organization.

*"The Company" is the term used throughout this paper to denote the hypothetical organization under investigation. The model has applicability to any medium-sized company with geographic dispersion of its production and operations.

A major task in any systems analysis is to formulate a problem statement or hypothesis to bound the analysis to be accomplished. The problem addressed by this analysis simply stated is: What is to be gained by moving to a computerized information system (CIS)? This question calls for an evaluation of the system (see Chapter 7). The primary criterion is the long-range success and stability of the Company. Historically the Company's single most important secondary criterion has been an adequate return on investment (ROI).

METHODOLOGY

The analytical approach will be the identification and development of a model to determine the secondary criteria that can be measured and that are considered for good reasons to be positively correlated to the primary criterion of long-range success and stability of the Company.

The model to be used is an information system model with multiple channels feeding data to various levels of the organization. The information needs of the Company are best shown metaphorically in Figure 16.2 as a pyramid with the apex being top management whose primary information needs are strategic. The Company information system must be designed to meet the needs at all levels. These needs can be further categorized as internal to the Company and external in the environment. Internal needs are: sales, production, inventory, personnel, financial, and forecasting. The external needs are linked to information concerning subcontracting vendors, competitors, location environment, and the federal, state, and local governments.

The final goal of all information systems is to aid in the decisionmaking process. To make decisions information must be valid and available at the time it is needed

Figure 16.2 The company information pyramid.

in the process. The span of perishability for operational data is very short, whereas the control and strategic data perishability spans a longer time period. Attempting to accomplish any of the functions of management with inaccurate or insufficient data can lead to system deterioration or failure (see Chapter 18).

ANALYSIS

Why should any organization consider the transition to a computerized information system (CIS)? Behavioral analysis reveals that the fundamental reasons for transferring to some form of mechanization in information systems are the same as any other facet of an enterprise. Classically companies turn to mechanization because (1) Company growth exceeds the capability to expand the present processes by any other reasonable scheme; (2) technology changes force different and more complex tasks that can best be mechanized; (3) efforts to cut costs force searches for alternatives; and (4) growth in data entities beyond manageable limits.

All of these reasons exist for information systems along with another highly important one. The decisionmaking process requires data from all elements of the enterprise within a diminishing time span as system complexity increases. This significant phenomenon forces an enterprise to consider CIS to meet its growth—and even survival—needs.

The advent of modern computing and communication devices makes economic feasibility less of a problem than in the 1960s or 1970s. Although cost remains an important variable—particularly for the smaller organization—the cost per unit of work drops in a CIS as growth occurs. For example, the cost per payroll check decreases within a CIS as personnel increases. Analysis reveals that CIS is an attractive option for an enterprise because it facilitates: (a) increased growth capacity; (b) assuming more complex tasks; (c) cost savings per unit of work; and (d) improved decision processes through better quality data being available more rapidly.

THE COMPANY INFORMATION SYSTEM

Analysis of the Company information system is necessary to assess the input of CIS on the system. The heart of the Company business is the manufacturing operation and this area has received extensive industrial engineering efforts in the past to improve operations. Hence, this area is chosen for analysis first.

Figure 16.3 depicts the Company's existing data systems in relationship to the manufacturing control system. This system is entirely manual at the present time with the exception of the payroll system, which is computerized through a local service bureau. Each of the manufacturing locations has a stand-alone manufacturing control system with all supporting systems performed at the corporate headquarters. The information flow in this sytem is complex with many feedback paths and is further complicated by having two manufacturing locations.

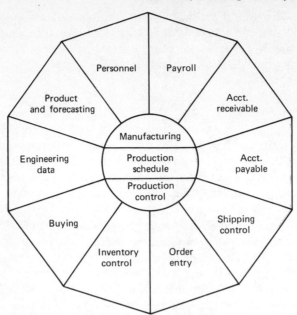

Figure 16.3 Corporate existing data system.

Figure 16.4 shows the information system data file locations. The order entry function initiates action and will be used as an analytical focus to illustrate some current problems. The current system relies heavily on mail service and phone orders from the four regional sales offices. The Company has Wide Area Telephone Service (WATS) so the most used order entry form is a telephone order followed by a mailed formal copy. This results in increased effort at the Corporate headquarters as the order is handled twice in the matching process when the formal order is received. All of the files involved are "tub" files organized by customer account and name. Daily order entries are accumulated to be placed in an order-in-process file organized by product number. The translation of the orders from customer name to product part number is an error-prone operation. Intensive control efforts have solved some of the error problems, but they do not relieve the saturation problem of peak loads occurring in the processing.

The company manufactures some 250 unique products with an average of 50 separate parts per product in any given production year. Orders for spare parts and replacement subassemblies is a major part of the order entry processing. The Company has some 4000 to 5000 active customers and some 10,000 inactive ones (less than one order per year). The order entry cycle alone requires 10 people at corporate headquarters and 16 at the regional offices. Telephone costs allocated to the order entry function typically are $5000 to $6000 per month which is one-quarter of the WATS charges for the Company. Total costs of the order entry function are $449,000 per year, or 0.5% of sales, which is an acceptable cost. However, adding capacity to the manual system requires more people and a linear

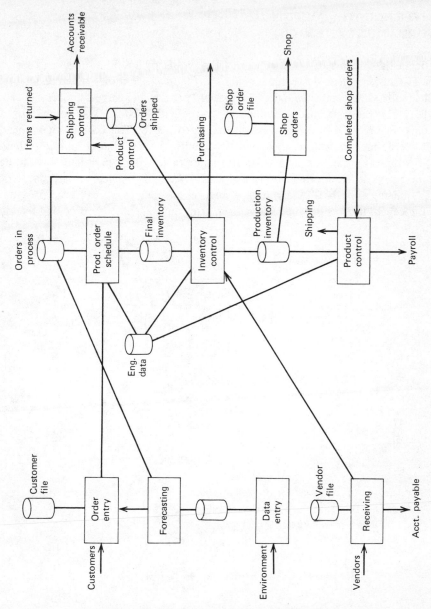

Figure 16.4 Information system data file points.

increase in the costs associated with this function while raising the probabilities of human error, lowered work quality, and customer satisfaction. Furthermore, the problem of peak workloads will continue.

The scope of this paper does not allow for analysis of all areas of the manufacturing control system but many of the weaknesses of the order entry function will be duplicated in other areas.

CONVERTING TO A COMPUTERIZED
INFORMATION SYSTEM (CIS)

To begin the process of moving to a CIS, a model of how the processing occurs in a Data Base oriented system is required. Figure 16.5 shows the fundamental model with which all corporate information processing can be conceptualized. Table 16.1 compares distinguishing features of the manual and computerized systems. Figure 16.4 is directly translatable into a CIS. The base files and interfaces are defined as well as those interfaces with the external corporate system.

It is proposed that each regional office enter their orders on a daily basis using a data entry device located at the regional office. This device would consist of a cathode ray tube (CRT) keyboard display connected to a small processor that has the capability to store the information, print out data, and transmit and receive

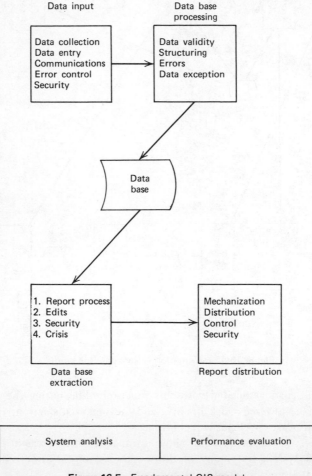

Figure 16.5 Fundamental CIS model.

Table 16.1 *Manual-Computerized Comparative Features*

Manual System	Computerized System
Application oriented	Data base oriented
Data distributed	Data centralized
Processes vary at different locations	Processes are more standardized
Data files are organized by application	Data files are organized by functions
Processing on some rigid schedule	Procession tends to be on demand
Difficult to accumulate data from different files	Data accumulation easier
Processing labor intensive	Processing not limited by labor; usually limited by machines. Error detection easier and earlier in the process
Data entry to the corporate system depends on mail and phone	Data entry more automatic and faster
Data records bulky historical records complex	Data records on magnetic media compact, easily stored off-site
Data security depends on people	Data may be encoded at entry and access limited from data base
Changes involve procedures and people	Changes involve software and less personnel
Cost of growth linear with personnel costs	Costs are quantized in levels machine and personnel costs
Training of new personnel easier	Training of new personnel more complex

data from a central site. The CRT would display an order entry form on the screen for the order clerk to follow. Savings will occur at the corporate headquarters once the CIS is operational due to being able to allow the regions to enter their orders directly without any help from corporate headquarters.

IMPACT ON THE COMPANY

With these relationships and potentials in mind the impact of CIS on corporate structure can be addressed. Historically, the controller division has supervised the data processing functions. There are many reasons for this; since this division collected data from all facets of the organization for financial reporting it seemed a natural place for data processing functions. Modern management theory has evolved the matrix organization that groups functional work units into groups that service the whole enterprise through project management. This concept allied to the Company organization would indicate a separate division for corporate information services such as diagrammed in Figure 16.6. This division would house the data processing and CIS design activities and serve as a skill center and career path for computer expertise for the Company.

The functions for each of the major elements of the CIS organization are shown. A staff function for overall system design, performance, and corporate liaison to determine needs is shown separated from line functions. Data Base Management is separated from the traditional Data Processing organization in view of its importance to the system. Likewise a Communications Services group is defined to handle all reports, data transmission, and Forms Design. The traditional data processing functions are, therefore, concentrated in one group. The CIS Director in the recommended new department would hold vice president rank due to the role of the CIS division, particularly with regard to its impact on corporate decisionmaking.

IMPLEMENTATION AND COSTS FOR CIS CONVERSION

This section presents a recommended plan to implement a corporate-wide CIS. This plan considers the necessary data processing hardware and software along with recommended staffing for all corporate elements. A 5-year schedule and estimated costs for operation of the CIS is presented.

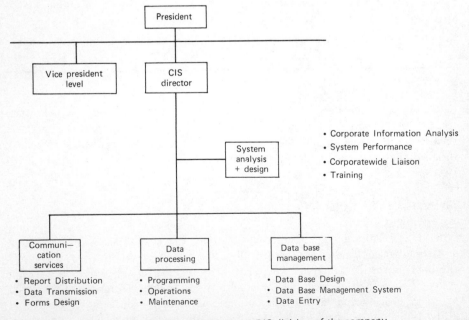

Figure 16.6 The proposed CIS division of the company.

Every effort has been made to utilize existing corporate facilities and personnel in these estimates. The attempt has been made to provide for future growth in configuration of the system and to provide a reliable secure operation.

This plan presents the overall framework for the CIS; however, the first activity of the plan calls for a more detailed investigation into facets of the CIS to be developed. Elements of the plan should include: (a) organization structure of the CIS department; (b) hardware configuration; (c) software configuration; (d) site preparation; (e) training; and (f) system analysis and design.

The corporate headquarters will house the central data processing facilities and the central data base. The central data processor communicates with the remote manufacturing site and the regional offices via the Company WATS telephone service lines to equipment for data entry and retrieval at these sites. The manufacturing site will have a remote job entry facility to allow execution of jobs and the development of software systems from this site. The data base module will be a software system that will allow the necessary access to the corporate files for both updating, creation, and retrieval functions. The proposed system will have the time-sharing module to allow both the on-site and remote-site soft-

Table 16.2 *CIS Conversion Cost Analysis (in Thousands of Dollars)*

	Years				
Item	1	2	3	4	5
Control Site					
Equipment	220	420	500k	550k	600
Personnel	415	615	670	700	750
Site Preparation	300				
Training	200				
Remote sites					
Manufacturing					
Equipment	—	84	100	110	120
Personnel	75	155	170	200	220
Site Preparation	100				
Training	100				
Regional					
Equipment	15	24	30	35	40
Personnel	200	220	260	290	330
Site Preparation	100				
Training	100				
	1825k	1518m	1730	1885	2060
Loading 50%	912	759	865	943	1030
Total	2737	2277	2595	2828	3090
Gross sales	100m	112m	134m	160m	193m
CIS as % of sales	2.7%	2%	1.9	1.7	1.6

ware development to have access to the system. The heart of the software configuration is the Operating System. This system is supplied by the computer hardware vendor, along with other optional software packages at additional cost.

Annual cost estimates for the CIS are projected in Table 16.2 and are based on the assumed hardware and software configuration. The first year includes some one-time startup costs such as site preparation and training. Costs are developed for the central site and the remote sites assuming that the CIS organization recommended is in effect and that the training for remote site personnel has been accomplished. A 5-year projection of CIS costs assuming a 50% corporate overhead factor on all costs is shown. Projection of CIS costs as a percentage of gross sales is shown as one measurement standard for management evaluation of the effectiveness of the CIS conversion program. Sales forecast show an annual 12% growth, which would be unlikely if management does not adopt this CIS proposal.

Although not developed here in detail the implementation schedule would show the last year being spent in specifications for the hardware and software, the information system elements, the site preparation and installation, and the acceptance of equipment. The emphasis on the second year would be in training of users and the installation of financial control systems for accounts receivable and payable; and the inventory control system. The third year would be spent finishing the primary manufacturing control systems and the final 2 years would be spent in development of auxiliary systems for purchasing and decision modeling. These models need a complete Data Base and hence should be deferred until this time.

RECOMMENDATION

The Company top management should adopt this computerized information system (CIS) conversion proposal.

17

Case Study in Hydroacoustic Systems: Hydroacoustic All-Weather Scoring System

FORTUNATO F. CATALDO

Mr. Fortunato F. Cataldo, at the time he accomplished this systems analysis in March, 1977, was Chief of Performance Assurance for Kentron International, Inc. stationed at Kwajalein Island, Trust Territory of the Pacific and a MSSM student.

Mr. Cataldo's analysis demonstrates:

1. Conceptualization of an improvement to an existing missile impact scoring system. (An example of Table 4.2 Systems Analysis level 9).
2. Quantitative analysis from acoustic physics and engineering with data reduction utilizing a Fortram IV computer program to support the finding of technological feasibility (see Chapters 6 and 8).
3. Normative research to support his contention that a hydroacoustic system would be preferable to the existing radar and hydrophone combination.
4. Costing analysis.
5. A Systems Analysis presenting potential improvement in operational efficiency which could trigger political feasibility for approval of the project.

R. M. KRONE

INTRODUCTION

This study investigates the feasibility of using an all-weather impact scoring system in the Kwajalein Lagoon. The optimum array diameter was selected based upon transmission losses found in shallow water coral lagoon environments. The shore station equipment configuration and mathematic model can accommodate multiple impacts into the hydrophone array. The shore station can also be unmanned; during an operation, a timing clock would turn on the shore station recording equipment. Data would be transferred to the central computer facilities on Kwajalein for data storage and recall for later mission analysis.

A cost effective study shows that the savings to be realized with an all-weather scoring system during a typical 2-year mission period would be substantially greater than the cost of installing an all-weather scoring system.

Requirements for measurements and recovery of intercontinental ballistic missile (ICBM) reentry vehicles (RVs) impacting in Kwajalein Lagoon were established for technical and ecological reasons: (1) impact time is important to the missile designer and reconstruction of trajectory and evaluation of warhead arming and fuzing subsystems, and (2) location of the impacting RV in azimuth and range is necessary so that the RV can be recovered for metallurgical examination and to preserve the ecological environment by removing RV debris from the lagoon bottom.

The present Hydroacoustic Impact Timing System (HITS), which obtains the required measurements at Kwajalein Missile Range, is shown in Figure 17.1. There are two independent systems: A modified Ka Band side looking radar, called the splash detection radar (SDR), and the hydroacoustic system. Together these systems provide data on azimuth and range and record impact time of RVs on the lagoon surface. The hydroacoustic system consists of a shore station and four concrete anchor blocks which support the hydrophones, velocimeters, and cable terminations. Three anchor blocks hold a hydrophone and velocimeter; the fourth anchor block holds a hydrophone and an acoustic projector. The lagoon location is shown in Figure 17.2 as well as the hydrophone locations of H_1, H_2, H_3 and H_4. The mathematical combination of near field velocimeter time and acoustic projector time, as received by the hydrophones, yields a nominal mean velocity of sound within the HITS coverage area.

All bottom-mounted equipment is tethered to the shore station by submarine cable. Electrical energy provided by the shore station power amplifiers is used to pulse the velocimeters at H_2, H_3, and H_4 to obtain a direct digital sound velocity reading at the shore station. The acoustic projector accepts the electrical energy, converts it into acoustical energy, and transmits it through the water. This energy is picked up by H_2, H_3, and H_4 hydrophones and sent via the submarine cable to the shore station for recording and computer program input to develop the nominal velocity of sound within the hydrophone array.

The SDR records the splash of the RV in time, azimuth and range, and vectors the diving barge to the impact area for recovery.

The Ka frequency of the SDR cannot penetrate rain, and without azimuth and range to the impact, the RV cannot be located. Because of this, missions are often delayed or cancelled because of inclement weather. The proposed system is an all-weather scoring system that will eliminate this costly problem.

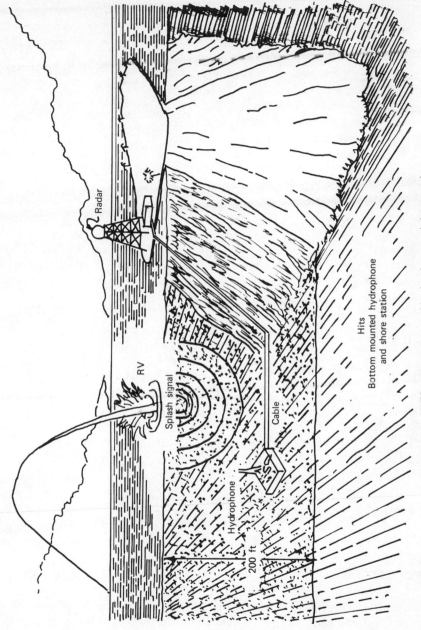

Figure 17.1 HITS bottom mounted hydrophone and shore station.

Radar

RV

Splash signal

Hydrophone

200 ft

Cable

Hits
Bottom mounted hydrophone
and shore station

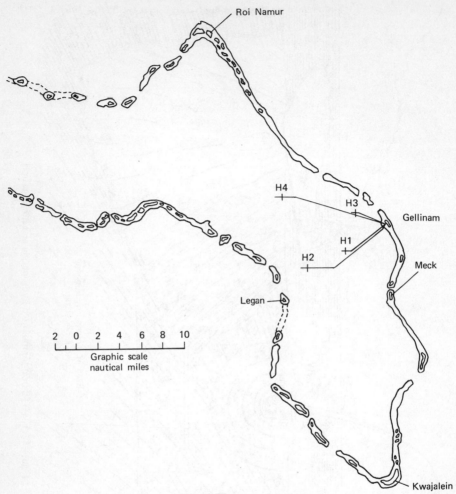

Figure 17.2   HITS system location.

BASIC ALL-WEATHER SCORING
SYSTEM CONSIDERATIONS

The proposed system would eliminate the need for the SDRs. RV impact would be located with the use of a hydrophone array. Consequently, the system would not be clear weather dependent; but would produce the data required even during inclement weather.

The hydrophone sensor (acoustic transducer) provides very little resolution capability since only time of occurrence is available. Therefore, these sensors are used in arrays. An array can provide the difference in arrival time of the impact signal at each hydrophone, which gives the relative distance for each element of the array from the impact point. These data can then be used mathematically to determine impact location when the location of each hydrophone and the velocity of sound and propagation path are known.

Acoustic systems are limited by the characteristics of the medium in which they operate. Acoustic characteristics will differ due to temperature and salinity of the water, and topography of the bottom. The RV impact area of Kwajalein Lagoon is a mean 200 feet in depth and falls within the operational domain of shallow water acoustics, which is what we shall consider for this analysis.

A theoretical study was made to determine the number and location of sensors required for an underwater acoustic system to monitor a 10-mile diameter area in the Kwajalein Lagoon. The signal present at a receiver is equal to the difference between the transmitted source level and the transmission losses. The resulting signal-to-noise ration (SNR) at the receiver must be adequate for detection and must be within the frequency bandwidth of the hydrophone.

When acoustic energy is transmitted in water, the propagation will be refracted from a straight line. Maximum range from a source to a receiver is seriously affected by the velocity profile. Zones of very low sound intensity (shadow zones) can exist. A computer-programmed raypath analysis was run using a velocity profile designed as being typical for the Kwajalein Lagoon to determine the adequate coverage.

Water depth was chosen as 200 feet, and a flat bottom assumed. The maximum range of a direct path propagation from the surface to the bottom was about 1.75 nautical miles for water depths of 200 feet or less. In the case of bottom reflections, amplitude of the reflected ray will increase from a finite value at normal incidence to a value of unity at angles equal to or greater than the critical angle. The critical angle (O_c) which is measured normal to the bottom, is given by

$$O_c = \sin^{-1} C_1 / C_z \qquad (17.1)$$

where C_1 and C_2 are the velocities of sound in water and bottom, respectively.

In the lagoon, which has for the most part a hard coral bottom, the critical angle should be less than 85°. Acoustic energy arriving at angles of 5° or less with respect to the bottom will be almost totally reflected. The large acoustic mismatch at the interface between the surface and the atmosphere will cause the amplitude of the surface reflection to be almost equal to that of the incident ray. Shallow water propagation of sound energy in the Kwajalein Lagoon over rock or coral bottoms is very good, and the decay of sound intensity with range is less than the normal inverse-square law.

Since the raypath analysis showed no shadow zones out to ranges in excess of 8 nautical miles, a hydrophone array could monitor a 10-mile diameter area. The array would consist of seven acoustic sensors, with six of these approximately evenly spaced around the circumference of the 10-mile diameter area, and the seventh sensor in the center (see Figure 17.3).

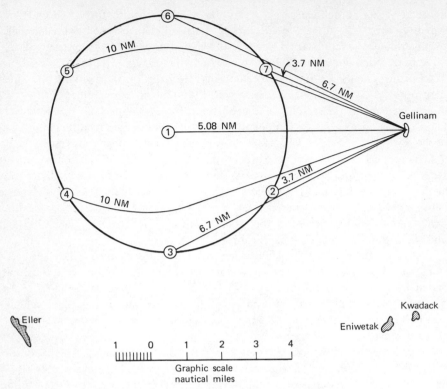

Figure 17.3 Proposed all-weather scoring hydrophone configuration.

TRANSMISSION LOSS

The transmission loss of the impact signal on any five of the seven hydrophones is required for solution of impact location. A tape recording of an RV impact which adequately described the impact signature characteristics, frequency, and signal level was used to determine the maximum transmission loss, based upon the following equation, and considering spherical spreading

$$L_M = 20 \log R + \alpha R \times 10^{-3} \qquad (17.2)$$

where

L_M = maximum transmission loss (dB)

R = range (yards)

α = attenuation coefficient (dB/kyd)

Using a conservative estimate for the frequency band of interest of $\alpha = 0.15$ dB/kyd, and $R = 8.67$ nautical miles, L maximum would be 87 dB.

Coupled with the transmission loss and a reduction of the overall impact level are the lagoon's ambient noise spectra and broadband noise levels. The data used to determine these parameters were based upon recorded measurements made of the noise of a ship as it moved in the lagoon; the noise levels measured are conservative compared to actual noise levels measured by RV impact. The broadband noise level was +13 dB ref to 1 μbar for our analysis.

An estimate of +110 dB ref 1 μbar at 1 yard for the broadband acoustic energy of an RV impact value was determined by mathematical analysis of several RV impacts in the lagoon, and this value is also a conservative value.

The minimum signal to noise ratio (SNR), using the preceding values for RV impact source levels (S_0), maximum transmission loss (L_M), and broadband ambient noise (N_T) is

$$SNR = S_0 - L_M - N_T = + 110 - 87 - 13 = +10 \text{ dB} \qquad (17.3)$$

The results of the foregoing analysis show the energy levels that impinge upon a hydrophone in an array are more than sufficient for RV impact location. The impact location of a single RV can easily be determined by time recording the signal received from several hydrophones and then using the time difference of arrivals at each of the hydrophones to determine the impact position.

A basic method of determining impact locations of one or more RVs impacting within an array will now be presented.

IMPACT LOCATION BY HYDROPHONE

Hydrophones can be arranged in an array to determine RV impact locations within or near the array by implanting each hydrophone in a known location and measuring the differences in arrival times of sound at the stations.

One method of determining impact time is as follows: The impact of a missile in the water generates acoustic energy waves, which are radiated spherically from the point of impact. Considering the simplified two-dimensional, two-station diagram shown in Figure 17.4, any line of constant time difference between the two hydrophones (H_1, H_2) is a hyperbola, described by an equation of the form

$$\frac{x^2}{a^2} - \frac{y^2}{b^2} = 1 \qquad (17.4)$$

where a and b are constants that describe the location of the vertices of the focal points of each hyperbola.

It will be noted that two stations do not provide sufficient information for the unique determination of an impact point location. Since two times yield one time difference line, then the intersection of two unique time difference lines indicates the occurrence of an event at that point. With three hydrophones, three time difference lines are available. In no possible case are only two time difference

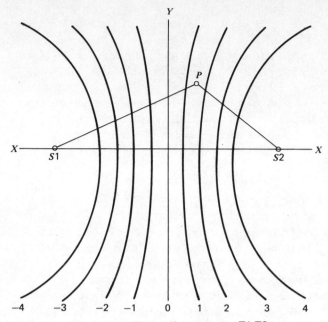

Figure 17.4 Time difference units, T1-T2.

lines available using all combinations of time from N number of hydrophones. Using two times, only one time difference exists. Using three times, three unique time differences exist: $T_1 - T_2$, $T_1 - T_3$, and $T_2 - T_3$. In general, the number of time differences (T_0) available from hydrophones (H) is governed by

$$T_0 = \frac{H!}{2(H-2)!} \tag{17.5}$$

where the differences are taken between two stations.

To determine a point based on these time differences, a third station is added, as shown in Figure 17.5. This system will actually provide a redundant solution for the two-dimensional case, since the intersection of any two of the three hyperbolae will uniquely determine the (X_p, Y_p) coordinates of point P, the RV impact point. The simple example for two dimensions can easily be extended to the three-dimensional case. The constant time difference hyperbolae become hyperbolae of revolution:

$$\frac{x^2}{a^2 - y^2} \frac{}{b^2 - z^2} b^2 = 1 \tag{17.6}$$

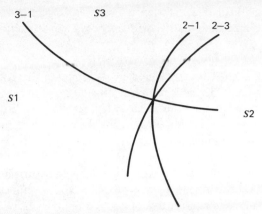

Figure 17.5 Determination of a point by the intersection of constant time difference lines for three stations.

Here, all three time differences from three stations will be required to find the (X_p, Y_p, Z_p) coordinates of the impact point. Three simultaneous equations of this form must be solved in order to determine the three coordinates of the impact point.

This simple case is, of course, complicated by various uncertainties, such as the variation of sound velocity, subsurface characteristics, and other factors. The solution "point" is actually a volume, in three dimensions, because of uncertainties in the measurement. These uncertainties are discussed in more detail later.

REQUIRED SYSTEM CONFIGURATION

Although only three stations are theoretically required for a triangulation solution for the three coordinates of an impact point, a practical system will also require the time of impact to be included in the solution. Each hydrophone in an array is located in a Cartesian coordinate system (X, Y, Z) with a unique set of coordinate values. Four time differences are required to solve four equations for the four unknowns X_1, Y_1, Z_1, and time of impact; four hydrophones will actually provide six time differences for stations recording the event. Although four stations could ideally provide the coordinates and time of impact, a practical hydrophone array requires more than this minimum number of stations for the following reasons:

1. The possibility exists for one of the hydrophones to be inoperative, or to fail to record the given event. In this case, no solution would be possible from a minimum array.
2. The quality of the data from one of the hydrophones might be poor so that the data was not usable.

3. A redundant mathematical solution is desirable to apply the method of least squares to optimize the solution with respect to the uncertainties in the measurements.

Another factor that must be considered in determining the configuration of the array is that it is very desirable to place one hydrophone as close as possible to the nominal impact point. This is advantageous because a good signal of the event is most likely at this location, minimizing the errors introduced in the system.

In summary, it is recommended that the array consist of a center hydrophone with six hydrophones arranged equally spaced in a circle centered on the center hydrophone. The minimum feasible array would be four surrounding a center hydrophone, but the area of redundant coverage would be undesirably small. The marginal cost of adding the fifth and sixth hydrophones is very small compared to the total system cost, and the reliability and quality of the data should certainly justify this incremental cost.

MATHEMATICAL DATA ANALYSIS

The basic data reduction requirement for the redundant hydrophone system described here is to determine the four unknown quantities, X_1, Y_1, Z_1, and T_1, which are the Cartesian coordinates and the time of the impact event from the known hydrophone locations, X_i, Y_i, Z_i, and T_i, the time of arrival of the signal at the ith hydrophone. The time (T) required for the signal to travel from the point of impact to any hydrophone is related to the distance (D) by the propagation velocity, V by the equation

$$D = VT \qquad (17.7)$$

Applying this equation to the ith hydrophone and the point of impact in Eq. 17.7 results in the relationship

$$[(X_I - X_i)^2 + (Y_I - Y_i)^2 + (Z_I - Z_i)^2]^{1/2} = V_i(T_i - T_I), \qquad 4 \leqslant i \leqslant 7 \ (17.8)$$

The subscript i, related to the hydrophone number, will vary from four, the minimum required for a solution, to seven, the maximum number of hydrophones. The propagation velocity V_i, for each hydrophone will be determined from pre- and postmission calibration data.

The number of hydrophones (N) required for a solution for four unknowns (r) can be stated as follows:

$$N = \frac{N!}{r! \, (N-r)!} \qquad (17.9)$$

which yields the following table or unique solutions for from four to seven hydrophones recording data:

N	Number of Unique Solutions
4	1
5	5
6	15
7	35

Since the number of unique solutions generally exceeds the number of unknowns, an overdetermined solution exists, and the method of least squares may be applied to optimize the solution for the unknown quantities. Each solution will represent a "point" in the four dimensions X, Y, Z, and time. The principle may be illustrated by the two-dimensional case illustrated in Figure 17.6. The problem is to locate the impact point (IP) in: (X_0, Y_0) such that the sum of the squared distance from this point to each other solution point (X_n, Y_n) is minimized. The squared distance is

$$d_j{}^2 = (X_j - X_0)^2 + (Y_j - Y_0)^2 \tag{17.10}$$

and the sum of the squared distances

$$\sum_{j=1}^{n} d_j{}^2 = \sum_{j=1}^{n} (X_j - X_0)^2 + \sum_{j=1}^{n} - (Y_i - Y_0)^2 \tag{17.11}$$

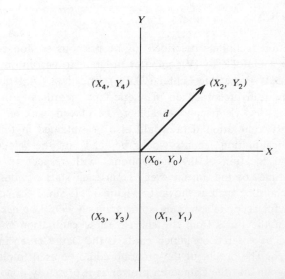

Figure 17.6   Illustration of the method of least squares in two dimensions.

Differentiation of this sum with respect to the quantities to be determined, X_0 and Y_0, yields

$$\frac{\delta\epsilon}{\delta\epsilon X_0} = -2 \sum (X_j - X_0)$$

$$\frac{\delta e}{\delta Y_0} = -2 \sum (Y_j - Y_0)$$

(17.12)

where the summation ranges have been left off for simplicity. The second partials are obviously positive, so that the first partials may be equated to zero for a minimum. This results in

$$X_0 = X_j/n \qquad Y_0 = Y_j/n$$

(17.13)

which is the arithmetic mean of the data points in each coordinate, where the summation ranges have been left off for simplicity. The extension of this to more than two variables is obvious. For the particular case discussed at the beginning of the section,

$$X_I = X_j/n \qquad Y_I = Y_j/n \qquad Z_I = Z_j/n \qquad T_I = T_j/n$$

(17.14)

are the optimum solutions for the four unknowns.

DATA REDUCTION

The data reduction technique described in the previous section could be implemented utilizing a FORTRAN IV computer program to be run on the CDC 7600 at the Central Data Processing Center (CDPC). The data from the hydrophones would be automatically recorded on magnetic tape recorders at the shore station and sent to the DCPC via telephone cable to be played back on an oscillograph recorder. Velocity calibration data would also be obtained in this manner. The times from the calibration records would be read and the velocities calculated from the known location of the hydrophones with respect to the calibration projector. The times of the impact event from each station would be manually read from the recording such as shown in Figure 17.7. Since correct time readings are very critical for accurate solutions, some training would be necessary for consistent reading of the times from the charts. These calibration and signal arrival time data would be entered on punch cards at the Data Center and the program could be executed. The output of the program could be used to place coordinates on a scoring diagram and the Camcomp plotter as is presently done with the other scoring systems.

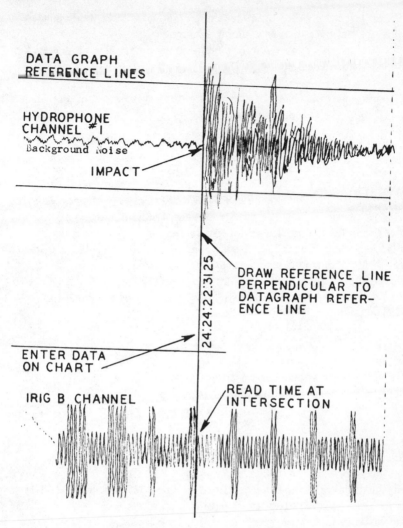

Figure 17.7 Typical hydrophone signal recording.

ERROR ANALYSIS

Earlier it was shown that the total system considers only three variables: time, distance, and velocity. Timing errors must be one of only two types: a nonsynchronized oscillograph chart or an inaccurate reading from an oscillograph chart. If the timing is incorrect by the same amount for each hydrophone, then this bias will be included in the solution only in the time of impact. One or more bad times will affect the total solution in X, Y, Z, and time of impact. The number of bad times, the magnitude of each bad time, and the total number of times used will each affect the solution. The magnitude of change is then a function of the input error. An inaccurate chart reading of time is identical to nonsynchronized time, since this, also, is an input error.

Impact location errors relative to the launch point may be divided into three major categories. These are:

1. Geodetic position error of the local reference relative to the launch point.
2. Errors in sensor position relative to the local reference.
3. Error in impact location measured relative to the sensor.

The accuracy desired in impact location is 100 ft or less, relative to the launch point. An obvious first step in the formulation of an impact location system is to establish the accuracy required in various system components. The local reference is Kwajalein, which has been surveyed. The atoll survey accuracy is 100 ft at a 90% confidence level. Position errors are commonly based upon the assumption of a circular probability distribution, in which case, 90% confidence corresponds to 2.15 times the standard deviation of error. That is σ 100/2.15 = 46.5 ft. This geodetic reference error provides the base upon which all system errors must be added. RMS impact location error, which is the total error relative to the launch point, is given by

$$\sigma_{PI} = \left(\sigma_{REF}^{-2} + \sigma_{SR}^{-2} + \sigma_{IS}^{-3} \right)^{\frac{1}{2}} \tag{17.15}$$

where σ_{REF} = rms geodetic uncertainty (error) of reference point, σ_{SR} = rms sensor (hydrophone) position error relative to reference, and σ_{IS} = rms impact location error relative to the sensor. The sensor system error, then, is defined by

$$\left(\sigma_{SR}^{-2} + \sigma_{IS}^{-2} \right)^{\frac{1}{2}} \tag{17.16}$$

The significance of the 100-ft accuracy goal in terms of system error depends upon the desired confidence level. For example, if a confidence level of 90% is desired, the corresponding sensor error is 25 ft rms.

The last possible error which must be considered is that for the velocity of sound in water. Willson's empirical equation is

$$V_z = 4422 + 11.25T - 0.045T^{-2} = 0.0182Z + 4.3\,(S - 34) \tag{17.17}$$

where V_z = velocity of sound in water (ft/sec) at z depth

Z = depth in ft from the surface

T = temperature ($^{\circ}$F) at Z depth

S = salinity in parts per thousand

Perhaps the least significant variable in the equation is the salinity factor. This averaged from 34.4 to 34.5 parts per thousand in the Kwajalein area.

With regard to temperature and depth, only the temperature contributes more to the velocity than the depth in shallow waters. At some depth, about 250 ft, a trade-off will be reached where depth then becomes the critical factor. This is true since the temperature then will remain "constant." The following table lists some typical values for V_z with $S = 35$:

Temperature, $F°$	V_z, Surface, ft/sec	V_z, 150 ft, ft/sec
65°	4967	4970
70°	4993	4996
75°	5017	5020
82°	5046	5049

Throughout the previous discussion where depth was mentioned, pressure was ignored. However, for the purpose of this study, pressure is synonymous with depth, and depth relative to hydrophone location is the factor of interest.

The velocity of propagation in water must also be considered since the medium is nonhomogeneous. With the use of Snell's Law, individual layers of water will be considered such that a corrected average velocity is associated with each hydrophone. The number of layers would, of course, be dependent upon hydrophone depth.

Considering the three error sources, time, distance, and velocity, the error model is established by relating these errors to position the basic variable of interest. The error model is therefore composed of the factors estimated as follows:

Time error: 0.015 sec
Location (distance) error: 25 feet
Velocity error: 20 ft/sec with calibration

This model considers only first-order effects and not higher order effects, which describe the interaction between the error terms, which will be small compared to the first-order effects.

Each of the error sources must be related to the basic variable of interest, distance (position) error; this is done from the basic equation, (7). Because of this relation, the error is not constant over the array but depends upon the impact location with respect to the array center. Assuming that an impact near the center of the array is unbiased, the error estimates in terms of position error result in the following position error estimates, due to the respective sources:

Error Source	Standard Error Estimate	Distance Relation	Position Error Estimate for $D = 1$ mile, ft
Time	0.015 sec	D/V	75
Distance	25 feet	1/1	20
Velocity	20 ft/sec	D/T	20

An estimated system error may be given by the following:

$$\text{Error}^2 = E_{\text{Time}}{}^2 + E_{\text{Distance}}{}^2 = E_{\text{Velocity}}{}^2 \qquad (17.18)$$
$$E_{RSS} = (E_T{}^2 + E_D{}^2 + E_V{}^2)^{\frac{1}{2}}$$

The estimated error for impacts within one mile of array center is therefore

$$E_{RSS} = (75^2 + 25^2 + 20^2)^{\frac{1}{2}} = 82 \text{ ft}$$

It is significant to note that the largest contributor to the error term is the time uncertainty; this is the reason for the recommended center hydrophone since the time signal is more likely to be "good" for this station due to its high probability of being near the impact point, resulting in a high signal-to-noise ratio and small error due to velocity variations affecting travel time.

Both the systematic and random error of a system deployed at KMR can be evaluated accurately and economically by conducting "splash" tests with the hydrophone system and the RADOT systems at Eniwetak, Omelek, Gellinam, and Legan. The splash events could be recorded by the hydrophone scoring system, the SDRs and the RADOTS. Triangulation from the RADOTS within the area of interest (described in the next section) results in errors less than 10 ft, so that the RADOTS could be used as a calibration standard for the hydrophone system and the SDR.

AREA OF COVERAGE REQUIRED

The area of coverage required for the KMR all-weather scoring system is determined by the characteristics of the expected target dispersion. For this study, the error distribution of the impacts will be considered to be circular, and it has been decided that an array of approximately 10 miles in diameter will be required to have a very high probability of impact within the hydrophone array. Classification prevents a detailed discussion in this analysis.

For the reasons stated previously, an array containing six hydrophones arranged in a circle around a center hydrophone is recommended. Figure 17.8 shows an idealized representation of the overlapping coverage from this arrangement, assuming a 10-mile diameter circle of deployment, with a radius of sensitivity of 5 miles. The numbers in the section indicate the overlapping coverage and illustrate how the redundancy varies with position.

HYDROPHONE LOCATION CONSIDERATIONS

Three main factors must be considered with regard to the location of a hydrophone. Probably the most important single factor is a clear, unobstructed view to the general area of interest; this is illustrated in Figure 17.9.

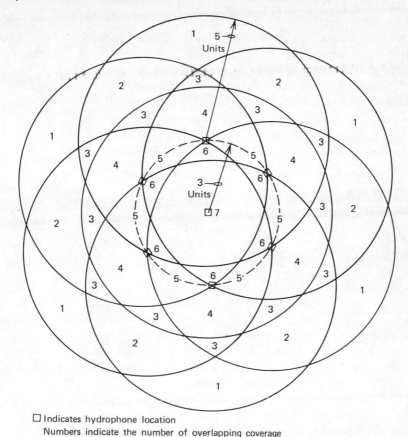

□ Indicates hydrophone location
Numbers indicate the number of overlapping coverage
circles for that sector

Figure 17.8 Illustration of coverage from an idealized hydrophone array.

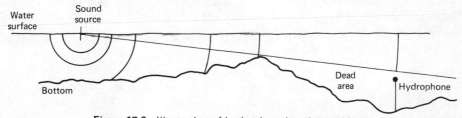

Figure 17.9 Illustration of hydrophone location problems.

The local environment for each hydrophone must be chosen to insure optimum coverage. Second, the depth of each hydrophone must be determined prior to installation, with regard to local terrain or bottom conditions. The third factor is the proximity to land-based hardware recorders. Obviously, the greater the distance to these recorders, the greater the cost of installation, repairs, and maintenance. As these factors decrease with a shorter distance, the accuracy for the exact location of a hydrophone increases. Transmission error decreases, but a trade-off must be made as the shoreline is approached since depth and local terrain factors still must be considered.

Because of the peculiarities of the particular situation at KMR, it will be impossible to strictly implement the system in a completely symmetric fashion. The location is actually constrained by the following factors:

1. Coral heads and other unevenness in the features of the lagoon floor.
2. Location on "smooth" floor area, for recovery considerations.
3. Location with respect to existing KMR target areas (classified).
4. Location with respect to inhabited areas, due to safety considerations for the target area.

PROPOSED SYSTEM DESCRIPTION

Although it is impossible to determine the optimum locations for the hydrophones without a detailed bottom survey, the Kwajalein Lagoon in the area of interest has been examined on a map that includes general subsurface features and lagoon depth. Figure 17-3 shows a typical layout that might be used, superimposed on this area. The nearest land to this area of interest is Gellinam Island. Cable lengths have been laid out, assuming the location as shown and utilizing Gellinam as a shore station for purposes of discussion.

A block diagram of the proposed system is shown in Figure 17.10. The proposed system would contain seven hydrophones, as described above, and a calibration projector, which would be controlled from the shore station. The electro-acoustic signals from the hydrophones would be transmitted to the shore station via the submarine cables and would terminate into buffer amplifiers that would provide the proper signal levels and impedance matching for interface with the tape recorder and/or oscillograph. A centralized control system would be used to operate the functions of all shore-based equipment. The projector is used for system testing and velocity calibrations before and after scoring missions. Figure 17.11 shows a typical shore equipment layout.

COST ESTIMATE

The total engineering, hardware, and installation cost of this system was estimated to be $1.9 million, including all hardware, surveys, and on-site logistic support. Exclusive of engineering cost and fees, the hardware and installation cost is distributed as follows:

Hydrophone and projector	$ 45K
Submarine cable (240K ft)	265K
Recorders (2)	65K

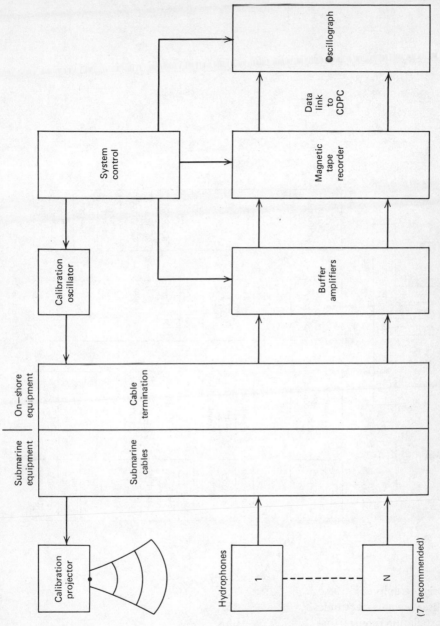

Figure 17.10 Proposed all-weather scoring system.

Oscillograph

Data link to CDPC

System control

Magnetic tape recorder

Calibration oscillator

Buffer amplifiers

On–shore equipment

Cable termination

Submarine equipment

Submarine cables

Calibration projector

Hydrophones

1

N

(7 Recommended)

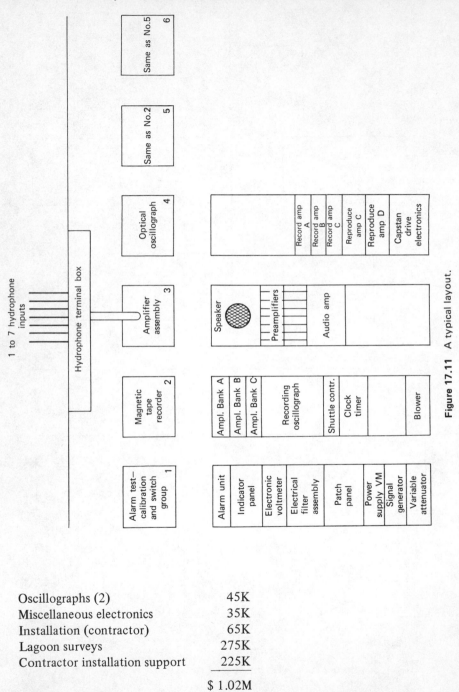

Figure 17.11 A typical layout.

Oscillographs (2)	45K
Miscellaneous electronics	35K
Installation (contractor)	65K
Lagoon surveys	275K
Contractor installation support	225K
	$ 1.02M

However, the system might be implemented for considerably less cost, assuming salvage of present cable and hydrophones. The following is essentially the equipment already at KMR for the Hydroacoustic Impact Timing System (HITS):

Recorders	(2)
Oscillographs	(2)
Hydrophones	(3)
Projector	(1)
Amplifiers	(1)
Test equipment	Miscellaneous

About 40K ft of submarine cable in several pieces, unknown condition

This equipment will be completely disassembled, and would require about 12 man-months of engineering and installation for reassembly. Assuming the above equipment could be used, and that the cable and hydrophones were available, the following "hard cost" estimate is made:

System	
Hydrophones (5), anchors, buoys, etc.	$ 60K
Submarine cable (275K ft)	350K
Miscellaneous electronics	25K
Subtotal	$ 435K
Support items	
Engineering, assembly, testing	60K
Survey (highly uncertain)	100K
Shore installation building	
(assuming addition to present facility)	100K
Installation support	150K
Software development	25K
Total estimated costs	$ 870K

The system would require the addition of about two manyears of effort to the Technical Support Contract for operation and maintenance of the system.

Although this study considered only one location in any great detail, the cost estimates could be projected to implementation for more than one target area. Since the cable length limitation is about 10 miles, the cost of the cable would be about the same for an installation at another location. However, the incremental cost for the second and third systems will be greater than for the first system, since the first system will exhaust all equipment available from HITS. The following table summarizes the projectee costs of installing one, two, and three All-Weather Scoring Systems:

Number of Systems	Incremental Cost, $M	Total Cost
1	$0.83M	$0.85M
2	$1.21M	$3.06M
3	$1.21M	$4.05M

The incremental cost estimate of $1.21 million for the second and third systems is based upon the original HITS hardware and installation cost from the first table above, allowing for about a 20% increase in prices.

SUMMARY

Present mission delays because of inclement weather over the range instrumented impact areas cost the range user $275K for a cancelled operation and approximately $68.75K for each hour a mission is delayed. Missions delayed or cancelled over a two-year period cost the range user approximately $2 million—more than the amount required to install one all-weather scoring system. There is no indication that present range requirements will be relaxed, and scoring and recovery will be necessary for years to come.

The cost for instrumentation for additional target areas could be eliminated by targeting those RVs that must be recovered into the proposed all-weather scoring area, and targeting all others into the broad ocean area (BOA), where recovery would be impossible with present recovery technology because of the ocean depth of 2500 fathoms.

BIBLIOGRAPHY

Urick J. Robert, *Principles of Underwater Sound for Engineers,* McGraw-Hill, New York (1967).

Camp W. Leon, Stern Richard, Brown M. B., *Underwater Acoustics,* Wiley-Interscience, New York (1970).

Caruthers W. Jerald, "Fundamentals of Marine Acoustics," Volume I and II, Sea Gant Publication No. TAMO–SG–71–403, College Station, Texas (1971).

"Proposal for a Balloon Mounted Impact Location System of Eniwetak," Bendix-Pacific Division, August 1965.

18

Case Study in Decision Capability: The Probabilistic Decision Display System

ROBERT KRONE

The capability for rapid provision of data, information, surveyed opinion and knowledge to decisionmakers has been made possible through computer and communication technology of the 1950s, 1960s, and 1970s. One method of linking that capability with the needs of policymakers of the 1980s will be the probabilistic decision display system (PDDS). This system will enable analysts to accomplish sensitivity and feasibility testing for decisionmakers of alternative assumptions, values, strategies, policies, and problem solutions on which the analysis and decisions are based. Decisionmakers, however defined, can ask for presentation on large conference room screens, or on home television sets, of the parameters of the analysis displayed along with the probabilities assessed of their accuracy and of their impacts on future reality. In this manner the time requirements for the search for alternative policies or the creation of as yet unidentified alternatives can be dramatically reduced over any previous analytical methodology. The highest payoff for a PDDS will be for crisis management situations; however, the techniques will have a wide spectrum of applications. The data requirements and systems analysis expertise of a PDDS will be high.

THE NEED

Management experts of the last three decades frequently assumed that good management would eliminate future organizational crises. Experience of that same time period has shown that assumption to be fallacious. Management did improve. However, if the definition of a crisis includes the threatened survival of the system or of a system subcomponent thus necessitating diversion of unplanned resources, the percentage of time that public and private organizations have been in a crisis management condition has also increased. My own informal surveys of American managers over the 1976 to 1979 period shows them estimating that their organiza-

171

tions were in a crisis mode through a range of 40 to 70% of the time. The place-ment of the judgement within that range depended upon the type of organization. Public systems seemed to be in a crisis mode more frequently than private ones and the closer the managers surveyed were to the operational functions of the system—as opposed to the policy and planning functions—the higher was the estimate of crisis conditions existing.

The reasons for this phenomenon appear to be in the increasing complexity and interactions of public and private organizations. Management and decision-making in a crisis demands higher knowledge and expertise which may not be available. The trends during the same period toward decreasing personnel and material resources aggravated the situation as crisis resolution demanded the best people and commitment of unprogrammed resources. This, in turn, contributed to generating more crisis situations for the future. Implications of this increasing trend toward crisis management are outside the scope of this study. Both existence and impacts of the phenomenon have not been adequately researched.

The probabalistic decision display system methodology has a high potential for assisting decisionmakers and managers involved in normal or crisis management over the next two decades. It will provide a major improvement for policymaking systems and a focus for development within policy sciences.

DESCRIPTION OF THE PROBABILISTIC
DECISION DISPLAY SYSTEM (PDDS)

Decision involves the future. Future predictions must entail estimates or probable error since uncertainty is involved. There is also uncertainty about the accuracy and sufficiency of past and current information on which decisions are based. Complete certainty that a particular event will occur in the future is equated with a probability of one (1). Assigning a probability of zero (0) means our assessment is that the event will never occur. Less than complete certainty is characterized by a probability somewhere between zero (0) and one (1). Table 8.2 and discussion of Chapter 8 present the fundamental quantitative decision model which is also the basic format for displaying the probabalistic decision data. The row titled, "Likelihood of States of the World" in Viewgraph 8-2 is where the probability figures computed for alternative future states would be portrayed. The PDDS would involve: (1) subjective probabilities (those based on the personal beliefs, feelings, or analysis of the persons who make the probability estimate); (2) relative frequency of occurrence (the percent of event occurrence over time under stable conditions or the observed relative frequency of an event in a large number of trials; (3) probabilities under statistical independence where the occurrence of one event has no effect on the probability of the occurrence of any other event; (4) joint probabilities under statistical independence (the probability of two or more independent events occurring together or in succession); and (5) conditional probabilities under statistical dependence when the probability of some event is dependent upon or affected by the occurrence of some other event. Bayes Theorem for revising prior estimates of probabilities based on new information (posterior probabilities) is an important computational tool for the PDDS as are applications of probability distribution concepts and those of unique isolated events.* Also relevant are my discussions of systems analysis methodology (Chapter 6), evaluation methodology (Chapter 7), qualitative variables (Chapter 5), and quantitative variables (Chapter 8).

THE DATA BASE

Data requirements and cost for a PDDS are large which partially accounts for the fact that none were in existence until the 1970s and none, to my knowledge, which meet all the specifications of this case study are in existence as of 1980. This discussion, therefore, is presenting a normative model for a probabalistic decision display system under the assumption that the computer and communications technology will allow economic and technological feasibility of such systems in the 1980s.† Data base requirements include: (1) all known behavioral research findings; (2) results of previous values analyses of individuals, groups, and entities involved or potentially involved in the problem under review (see Chapters 5 and 6); (3) a Talent Bank containing sources of knowledge from research organizations, other information systems or Delphi surveys††; (4) an ensemble of potential secondary criteria and standards for possible use in the evaluation process (Chapter 7); and (5) all alternative policies identified to date on the subject plus impact analyses accomplished on those alternatives.

ANALYTICAL AND DISPLAY
CAPABILITIES OF THE PDDS

Inquiries to the PDDS data base could be made for decisionmakers by analysts familiar with the system. Output displays of words, graphs, charts, comparisons, statistics, pictures, drawings, trend extrapolations, mathematical or economic formulas, alternatives and impacts would appear on large screens for face-to-face type conferences or could be transmitted to any televison or cathode ray tube receiver for electronic meetings of geographically dispersed participants.†††

*See Richard I. Levin and Charles A. Kirkpatrick, *Quantitative Approaches to Management*, 4th Ed. (New York: McGraw-Hill Book Company, 1978), Chapters 2 and 3; and, C. West Churchman, Leonard Auerback, and Simcha Sadan, *Thinking for Decisions: Deductive Quantitative Methods* (Palo Alto: Science Research Associates, Inc., 1975), Chapter 5.

†See Joseph P. Martino, "Telecommunications in the Year 2000," *The Futurist* (April 1979), pp. 95-103.

††For the Delphi Method of surveying expert opinion see Harold A. Linstone and Murray Turoff, eds., *The Delphi Method: Techniques and Applications* (Reading, Mass, Addison-Wesley Publishing Co., 1975). For a description of the Congressional Talent Bank of Futurists established in 1976 and used by committees and sub-committees of the United States Congress see Anna W. Cheatham, "Helping Congress to Cope with Tomorrow," *The Futurist* (April 1978), pp. 113-115.

†††See Robert Johansen, Jacques Vallee, and Kathleen Spangler, *Electronic Meetings: Alternatives and Social Choices* (Reading, Mass.: Addison-Wesley Publishing Co., 1979).

The probabilities associated with the economic, technological and political feasibilities of alternative policies (Chapter 6) could be presented for decision-makers who could also request the degree of credibility placed in various data and assumptions utilized by differing individuals or groups. For example, if it were a national level decisionmaking session concerning a civil-military instability in a neighboring nation the credibility of the intelligence information (probabilities of its accuracy as assessed by knowledgable parties) could be presented with the known facts at the time. If it were a Nuclear Regulatory Commission emergency meeting over an accident at a nuclear power plant (such as happened at the Three Mile Island, Pennsylvania plant in March 1979), the data base could be tasked to help sort out what was behavioral research data (what exists) versus what was values data (what various interest groups preferred) and what was normative data (views of what should be done). Impact predictions of alternative actions and the probabilities assessed concerning the accuracy of the information and the costs, benefits, and risk associated with each alternative policy could also be accomplished.

ILLUSTRATIVE USES OF THE PDDS

Applications of the PDDS are limited only by imagination. It would provide the optimum method for the systems analysis briefing as well as the method for bringing results of systems analysis to the public to increase the policy contribution capacity of citizens as discussed in Chapter 6. PDDS could be linked with "MINERVA" (Multiple Input Network for Evaluating Reactions, Votes, and Attitudes), a concept developed and tested by Amitai Etzioni and colleagues at the Center for Policy Research in New York. MINERVA would be a system of social communication by which large groups of citizens dispersed across the country could respond regularly to social, economic, and political situations.* Another compatible linkage would be William W. Simmons' invention of the Consensor—an electronic system providing participants at a meeting with terminals enabling them to express anonymously their degree of agreement or disagreement with a proposal along with a self-evaluation of their own expertise level. In Consensor results are immediately computed and presented graphically on a screen for an instant opinion poll of the group.†

For international security issues the PDDS could provide both early warning and crisis decisionmaking capability for serious events such as the Cuban missile crisis of October 1962; the August 20, 1968 Warsaw Pact move into Czechoslovakia; the fall of Saigon in 1975; the 1976 murder of two American Army Officers by

---

*Eugene Leonard, Amitai Etzioni, et al., "MINERVA: A Participatory Technology System," (New York: Center for Policy Research, Inc., undated monograph provided the author in June 1974).

†William W. Simmons, "The Consensor: A New Tool for Decision-Makers," *The Futurist* (April 1979), pp. 91–94. The article includes some projected future refinements that could move this system in the direction of the Probabalistic Display System presented here.

North Korean soldiers in the Joint Security Area of the Korean Demilitarized Zone,* the revolution in Iran of 1979 or the USSR invasion of Afghanistan in December 1979.

For purely military command decisionmaking the PDDS would be the optimum heart of any C^3I (Command, Control, Communications, and Intelligence) system.

In the Republic of Korea the governmental planning process reaches a milestone event on the first week of every year when the President is given a 30 min briefing by each of his Cabinet Ministers and central government agency heads. The purpose of the briefings is for the President to review how the previous years' programs have been implemented and the plans devised to implement national policies for the next year.† PDDS for this sort of national development progress and projection briefing would be ideal although political feasibility constraints would play a role (Chapters 5 and 6).

The PDDS would provide the appropriate analytical vehicle for multinational corporation chief executive officers to decide on corporate strategy, diversification or divestment plans or technology transfer ventures. The data base would attempt to meet the knowledge prerequisites for technology transfer provided in Chapter 12.

For conferences in any professional discipline the PDDS would provide a welcome replacement for the archaic tradition of reading papers and utilize, during the conference, the combined expertise of delegates to increase the knowledge base of the discipline. The PDDS would also offer real-time dissemination possibilities for the concepts and ideas brought to the conference.

For political campaigns PDDS would be a valuable assist for politicans to identify serious societal issues, to observe voter values toward those issues, and to provide constituents with alternatives resulting from analysis. This would accomplish a double-edged improvement of analytical skills of politicians and critical abilities of voters.

FUTURE IMPLICATIONS OF THE PDDS

The applications of a probabalistic decision display system are infinite. In general, the PDDS would be an important Policy Sciences tool for increasing the capacity of future public and private policymaking systems to achieve goals as well as to reduce uncertainties in the selection of goals and policies. This would raise the quality of policymaking in business, government, the military, academia, and in international entities. In crises, top decisionmakers often want assurances of certainty prior to making critical decisions, yet our complex world is full of uncer-

*For an analysis of the crisis communications and mechanisms of coordination used by the U. S. government in this case and the May 12, 1975 Cambodian attack on the U. S. merchant vessel "Mayaguez" see Richard G. Head, Frisco W. Short, and Robert C. McFarlane, *Crisis Resolution* (Boulder, CO: Westview Press, 1978).
†The Korea Times, January 22, 1978, p. 2.

tainties and probabilities.* More scientific systems analysis is essential to provide decisionmakers the best knowledge available concerning the risks, impacts and probabilities associated with converting that knowledge into improvement of the human condition. A commitment to this effort by some percentage of policy scientists will be required in the 1980s.

EPILOGUE TO CHAPTER 18 AND TO PART II, "PRACTICE"

To help in avoiding the hubris of scientism it will be well to remember the 16th Century—but still true—words of Sir Francis Bacon, Viscount of St. Albans (1560-1626), on the subject of studies and experience. In his essay, "On Studies," he said:

> To spend too much time in Studies, is Sloth;
> to use them too much for Ornament, is Affectation;
> to make Judgment wholly by their Rules, is the Humor of a Scholar.
> They perfect Nature, and are perfected by Experience.

*Bacon's Essays***

*Herbert A. Simon, the 1978 Nobel laureate in economics and a pioneer in the field of decisionmaking research, sees artificial intelligence produced through computer technology as a future method of assisting managers with their scarce resource of "attention" and to focus on quality aspects rather than quantity aspects of the data. See John M. Roach, "Simon says . . . Decision Making is a 'Satisficing' Experience," *Management Review* (American Management Association, January 1978), pp. 8-17. See also Herbert A. Simon, *The Science of the Artificial* (Cambridge, MA: M.I.T. Press, 1969).

**Bacon's Essays*, Sydney Edition, edited by Sydney Humphries (London: Adam & Charles Black, 1912).

Part III
SYNTHESIS

- Synthesis in systems analysis is an art practiced within a scientific paradigm
- Policy without theory is not defensible
- Theory without practice lacks validation

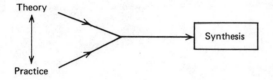

19

The Role of
the Systems Analyst

The role of the systems analyst is to synthesize: theory with practice; qualitative with quantitative tools; behavioral, values, and normative research with feasibility analysis; and to creatively alter or remove constraints to systems analysis. There are personal values implications for the analyst. The national and international role of the systems analyst in both public and private sectors is growing.

Analysis is the process of breaking down a complex system problem into understandable and manageable components. Synthesis involves the reassembly of those components into a new whole, which can more nearly achieve the aspirations of people. Systems analysis does both.

Systems analysis problems are all unique but there are usually portions of those problems which exhibit similar characteristics with portions of other problems elsewhere. Distinguishing the unique from the common or universal is a requirement of systems analysis. We err in applying tools previously found appropriate for a set of problems to the unique aspects of the problem at hand. American motivation methods to increase individual achievement and recognition may be counterproductive in a Japanese plant where the culturally based concept of "the company as a family unit" prevails. The future of aviation needs of the Hawaiian Islands are unique from anywhere else in the world. A corporate structure for the 1970s may be anachronistic for the 1980s and 1990s in view of the impact of international technology transfer and the ability to diffuse knowledge through automated management information systems. Soil penetrating fertilization will have unique benefits.

The role of the systems analyst is to understand and describe the problem, in its milieu, for himself and others and to increase the probabilities of success through the conceptualization of alternative models for the future.

In summarizing the role of the systems analyst for reasons of synthesis I would list the following nine major knowledge criteria;

1. Knowledge to evaluate existing and alternative future systems through a selection of secondary criteria, which demonstrably meet the prerequisites of secondary criteria—which are ". . . for good good reasons to be positively correlated

with, and more measurable than, the primary criterion of the system (e.g., security, improved health, effectiveness, economic development, success in business venture, better communication." See Chapter 7.

2. Knowledge to select and/or create the right systems analysis qualitative and quantitative tools for behavioral, values and normative research. See Chapters 5, 6 and 8.

3. Knowledge to deal more scientifically with constraints that inhibit or prevent total systems analysis. Figure 19.1 displays the seven major constraints facing the systems analyst of values, time, resources, distance, knowledge, tools and techniques, and political feasibility. The analyst can make major improvement contributions to systems through analysis which (a) identifies which constraints are functioning; (b) examines why they function and how they function; (c) creatively expands alternatives by altering or removing those constraints or including the time dimension in the analysis to do so in the future; and (d) provides normative models or guidelines to accomplish system goals more effectively and efficiently where constraints are fixed.

4. Knowledge to make explicit those methodologies used that lead to findings and conclusions.

5. System tacit knowledge sensitive to the problem identification, the techniques used, the findings obtained, and the recommendations given to policymakers. This should include an understanding by analysts of the role played by their own tacit knowledge.

6. Knowledge to understand the distinction between theory and practice while at the same time synthesizing them for systems analysis. The application of systems analysis to real-world problems is within both a scientific paradigm and a theoretical framework as well as for systems improvements that can be

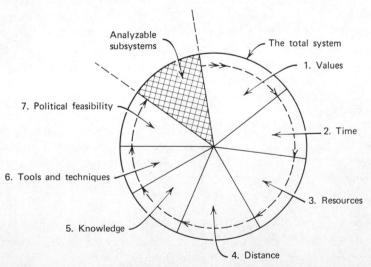

Figure 19.1 Major constraints that inhibit or prevent total systems analysis.

subsequently validated. When analysts use tools of an innovative or experimental nature it is encumbent on them to advise the policymakers of that fact. The medical metaphor has some relevance here. There must be a distinction between the application of medically validated techniques to improve the patient's health and the application of experimental treatment designed to improve the art and science of medicine. The medical metaphor has its limits when transferring to systems analysis, but the principle remains an important addition to knowledge criteria when conceptualizing the role of the systems analyst.

7. Knowledge of the linkages between systems analysis and political processes. Both political credibility and capability are highly dependent upon decision-making skills. Systems analysis has a role to play in improving the individual politicians as well as in assisting them through better tools and studies. Conversely, systems analysis has a role to play in providing the voting public more scientifically based analyses of public issues. When the complexity of issues exceeds the ability of the public to absorb those complexities democratic processes will deteriorate.

8. Knowledge to contribute to the on-going examination and improvement of systems analysis as a macro social tool. Systems managers may well be emerging as the most fundamental independent variable for impact on the dependent variable of quality and stability in human systems.[46] They are the focal point for the identification, analysis, and intelligent manipulation of other internal variables and responses to external variables. This involves management of knowledge as well as of people and technology. Systems analysis is the macro tool for the systems manager.

9. Knowledge of self values to answer the difficult question of "Whose values should I serve in this systems analysis?" when values conflicts occur. This knowledge involves wisdom and the inner synthesis of the analyst.[47] Managers, decision-makers, clients, workers, adversaries and analysts will bring varying perspectives—with associated values sets—to bear on the problem. Your ability to make a personal commitment of your life and work to a particular human system, or to a systems analysis, has a great deal to do with your personal values set. So does your degree of satisfaction or discomfort with your commitment to that system and its values. The exercise of Table 7.4 is designed to create an appreciation for this alternative values problem. After all, if you cannot completely follow Polonius' advice to ". . . above all—to thine own self be true, . . ."[48] you should at least know whose values you are serving.

PATHOLOGICAL SYSTEMS

Must the role of the systems analyst always be to help the system? What if the system itself is pathological? I need to address this bias of the book as quite obviously some systems should be terminated. Good systems analysis will identify termination as an alternative along with others for improvement or redesign. Knowledge for orderly termination of systems that have outlived their usefulness, and for systems to replace them—if necessary—should be included in the role of systems analysts.

Violent system termination falls outside the realm of science and into that of radical political-military action. An assumption of this book is that the systems analyst and decisionmaker have values consonant enough with the system to commit their expertise, their talents and their emotional involvement to the system's survival or improvement.

A dedication to the violent destruction of systems represents a values orientation toward terrorism, subversion, insurrection, or revolution. The failure of American strategic analysis to adequately consider the possibilities of the emergence of destruction-oriented groups, organizations, or nation states was motivation for Yehezkel Dror's 1971 book—*Crazy States*.[49] Since then the realities of what most members of the international community would call crazy actions have increased considerably to the degree that counterterrorism strategies are, at this time, receiving priority attention in many capitals of the world. My view is that systems analysts should dedicate their efforts toward those institutions, organizations, and governments that have the highest probabilities of improving the quality of life for peoples of the world and not toward values systems that postulate human suffering, assassination, cruel uses of innocent hostages, or genocide as appropriate means to achieve ends. I would, therefore, hope that systems analysts, systems managers, and policy scientists would not find their own values consonant with movements such as 11th, 12th, and 13th century Christian Crusades, German Naziism in the Third Reich of the 1930s and 1940s, or the terrorism of the Red Brigade. However, the possibilities of policy sciences and systems analysis knowledge, like technological knowledge, being used throughout the political ideological spectrum—for evil as well as good—are real, and deserve attention.

A goal of systems analysis is to improve the capabilities of the tools to the point where those perceptions of need for violent, destructive, and wasteful movements can be eradicated, but that goal appears too idealistic for the 20th century and relies on many social, economic, and political variables still outside the effective scope of systems analysis or policy sciences.

A final observation on the role of the systems analyst. Secretary of Defense Robert McNamara spurred the institutionalization of systems analysis in public policymaking in the U.S. Department of Defense with a staff of 13 analysts in 1961.[50] Today, 1980, there are systems analysts at every level of American local, state, and Federal government, at public and private research institutions, at Universities and in almost every country of the world. In January 1978 President Carter inaugurated zero-based budgeting (ZBB) as an add-on to the Planning, Programming, and Budgeting System (PPBS) of the U.S. Federal departments and agencies. The first indications were that analysis requirements inherent in ZBB might double the number of systems analysts required in the U.S. government and military services if implemented as envisioned. The technology advancements in private business in the 1970s and projected for national and international business in the 1980s are producing a corresponding demand for systems analysts in the private sector.

The role of the systems analyst is increasing.

20

Analysis and
Complex Problems

Analysis becomes complex precisely at those points where decisions have the greatest impact. Systems concepts and approaches are the best tools available for analyzing and synthesizing complexity but the dilemma of complexity remains. Those tools are still incapable of providing a complete understanding of the whole system. Systems analysis tools for handling complexity of the whole, of the parts, and of interaction of systems are improving. The awareness of the cost of failures as a major stimulus for sustained efforts in that direction, as is the phenomenon of exponential growth of knowledge and the need to improve linkages of knowledge to policymaking. The pitfall of real world distortion by the system scientist, as well as the atomistic scientist, is presented. Reasonable expectations of systems analysis are provided. The synthesis process for systems analysis has just begun.

Complexity in human systems is the phenomenon of multiple interrelationships and dependencies among people, structures, processes, and technology over time. Dealing with complexity is frustrating and perplexing but the ability to analyze complexity is also the skill with the highest payoff for the systems analyst and for the systems they analyze.

The problem of how to conceptualize technology transfer knowledge requirements (Chapter 12) exemplifies the complexity of modern issues for systems analysis. There is a long and growing list of other such issues beginning with the Club of Rome "five basic factors that determine, and therefore, ultimately limit, growth on this planet."[51] Those five were populations, agricultural production, natural resources, industrial production, and pollution. Some portions of those factors have been addressed in the case studies of Part II. But beyond those there are issues of communications, transportation, housing, public services, energy, weather control, genetics, conflict resolution, health care, human rights, security, food distribution, and resource allocation—all exhibiting increasing complexity for the decisionmakers, analysts, managers, and clients of human systems.

With this increase of complexity comes a related increasing cost of human system failure. Consider the cost of failure to find a new viable world economic order or the ubiquitous example, since 1946, of the costs involved should national governments fail to prevent nuclear war. Awareness of the impacts of human systems failures will continue to be the major impetus for the increasing use of systems analysis and policy sciences. This book has not pursued failure analysis in human systems but research superimposing the theory presented here over historical cases of human systems failures could add to our knowledge base.

Compounding the complexity issue is the knowledge that policies made and actions taken to alleviate problems may be creating those very same problems or more serious ones and resulting in a deteriorating spiral. This is what Jay Forrester has termed the "counterintuitive behavior of social systems."[52] His research showed a spiral of urban low-income housing programs causing higher population density that led to lower employment and lower incomes thus stimulating construction of more low-income housing.

One by-product of the systems approach is to produce a sense of humility in the face of knowledge available and knowledge yet to be discovered.[53] Biomedical knowledge has increased so much over the past 30 years that experts in the field cannot possibly absorb it much less be able to focus the totality toward their research.[54] As relying on curiosity to successfully handle knowledge diffusion has not worked too well, one of the major problem areas for Policy Sciences will be that of discovering new mechanisms for bringing available knowledge to bear on policymaking (Chapter 18 addresses that subject). For if we extrapolate the rate of knowledge growth of the past 30 years—doubling every 10 years—to the next 100 years, it means that our knowledge available today will be only 3% of that available in the year 2079.[55] The synthesis process for systems analysis is in its infancy.

For that reason synthesis does not mean a final solution. There are few permanent solutions resulting from systems analysis.[56] However, there should be more scientifically based and thoroughly considered alternatives developed within a better explicated values set. If achieved, that will produce a higher probability of the selected alternative achieving goals of those controlling the destiny of human systems as well as having those goals, and the means used to reach them, acceptable to the clients of the system—however defined.

For those who feel that systems analysis is establishmentarian, too expensive, or a dangerous tool of manipulators who know the utility costs and benefits but not the values I stand my ground on the statement that systems analysis has worked, is working, and will inevitably increase its areas of usefulness throughout the world. The issue is not whether systems analysis and policy sciences work, just as the issue is not whether nuclear fission works. The issue is to make it work effectively to achieve goals selected with wisdom.

C. West Churchman has stated his view that: "every world view is terribly restricted," and that systems scientists are just as prone to self-deception as the scientist employing the atomistic approach.[57] I prefer the word "distortion" to "deception," but agree that the pitfall is there for both specialists and systems scientists but contend that the latter should be more aware of the distortion wrought by their theories and models. Systems analysis can reduce the distortion of atomistic approaches to problem solving while instilling an appreciation set for the rational, extrarational, cultural, political and values variables involved. But Churchman is right that we cannot fully understand the complexities of today's world. To assume that we can would be deception. In spite of that, appreciation for the whole system sensitizes actors to other points of view, forces more explicit analysis of values and opens our justifications of feasibility to our own, and others', criticism.

As I was writing this final chapter's first draft President Anwar Sadat of Egypt, Prime Minister Menachem Begin of Israel, and President Jimmy Carter of the United States were together at Camp David, Maryland in an historic meeting searching for a peace formula for the Middle East. Those three national leaders will never "fully understand" each other or all the complexities existing. However, the new appreciation set of each of them which resulted from those 5 days of meetings were sufficient to achieve significant redesign of the human system we call the Middle East. Full understanding of complex systems is too much to expect in the 20th century. Perhaps when "Reality 21" arrives. Not now.

What then can we reasonably expect from Systems Analysis in the remainder of the 20th century? We can expect human systems improvements significant for ourselves, our families, our businesses, our governments, and our international social, economic, and political entities—certainly for the short term as measured against existing standards and, perhaps, for the long term as systems analysis is also a tool for evaluating existing standards and helping in the formulation of more relevant ones for the future.

Arnold Toynbee, after a lifetime devoted to the study of history, asked the question, "What is the lesson of our unforeseen and disillusioning experience?" He answered that: "It has taught us that, although progress may be cumulative in the fields of science and technology, there is no such thing as cumulative progress toward an ever-greater humanity in our treatment of each other."[58] Systems analysis, conceived within a policy sciences paradigm is an instrument for improvement in science, in technology, in management, and in the quality of the human condition. As such it will also become an important component in developing for future needs what Yehezkel Dror—in the Foreward to this book—has termed "a novel decision-culture".

Part IV

CHAPTER NOTES, REFERENCES AND BIBLIOGRAPHIC ESSAY

Systems Analysis and Policy Sciences is designed to be heuristic; to lead students and systems analysts to their own discoveries within an expanded view of systems analysis theoretical concepts. The technique of utilizing tables that condense large numbers of complex concepts to a single audio-visual two-dimensional medium of necessity abstracts the relevant literature and often presents conclusions in the absence of methodical and logical development and defense of those concepts.

Part IV is provided for those readers who wish to explore the relevant literature to satisfy themselves concerning the scientific basis of those concepts. The Bibliographic Essay provided for each chapter of the theory portions of the book (Chapters 1–8, 19, and 20) is not intended to contain an exhaustive listing of sources but rather significant works from which the reader can challenge or pursue in greater depth the ideas presented in this volume. With the recent computerization of bibliographic sources the reader should consider this listing as only a heuristic departure point for in-depth investigation of any knowledge areas touched upon. References provided through footnotes are not repeated in the Bibliographic Essay and entries are not duplicated, even though applicable to more than one chapter.

CHAPTER 1, KNOWLEDGE AND SCIENCE

Chapter Notes and References

1. Yehezkel Dror calls this "metacontrol knowledge, that is, knowledge of the design and operation of the control system itself." See his *Ventures in Policy Sciences; Concepts and Applications* (New York: American Elsevier, 1971), pg. 3

for a definition and pp. 9–24 for his explanation of the need for control knowledge and policy sciences.

2. During the 20 year period of 1946 and 1966 Michael Polyani wrote a stream of books and articles developing, inter alia, his theory of personal knowledge which I consider a fundamental scientific basis for my contention that systems analysis, to be relevant, must be able to incorporate both rational and extrarational components. See particularly his: *Science, Faith and Society* (Chicago: University of Chicago Press, 1946; New York: Oxford University Press, 1946, and Phoenix edition, Chicago, 1946); *Personal Knowledge* (Chicago: University of Chicago Press, 1958; London: Rutledge, 1958; and New York: Harper Textbooks, 1964); *The Study of Man* (Chicago: University of Chicago Press, 1958; London: Rutledge, 1958; Phoenix edition, Chicago, 1964); and *The Tacit Dimension* (New York: Doubleday & Co., 1966; Anchor Books edition, 1967).

3. I should qualify this statement with "for the short term future of humanity." The physicist Gibbs and the mathematician Wiener may well be correct in their concepts of entropy in the universe which predestines mankind to a universe in which chaos is most probable and order is least probable; where the characteristic tendency of entropy is to increase; and where "As entropy increases, the universe, and all closed systems in the universe, tend naturally to deteriorate and lose their distinctiveness, to move from the least to the most probable state, from a state of organization and differentiation in which distinctions and forms exist, to a state of chaos and sameness." See Norbert Wiener, *The Human Use of Human Beings: Cybernetics and Society* (New York: Doubleday Anchor Books, 2nd ed. revised, 1954; originally published by Houghton Mifflin Co., 1950). For *Systems Analysis and Policy Sciences* I am content to consider human systems' futures of 100 years or less.

4. Two good references are: Ernest Nagel, *The Structure of Science: Problems in the Logic of Scientific Explanation* (New York: Harcourt, Brace & World, 1961); and Abraham Kaplan, *The Conduct of Inquiry: Methodology for Behavioral Science* (San Francisco: Chandler Publishing Co. 1964).

5. Thomas S. Kuhn. *The Structure of Scientific Revolutions,* 2nd ed., (Chicago: University of Chicago Press, 1970), 1st edition 1962. This book is a modern classic which has, itself, achieved the status of a "scientific paradigm" by presenting a general accepted world view of evolution and revolution in scientific discovery.

Bibliographic Essay

An excellent essay on the distance and the linkages of traditional academic disciplines to the applied social sciences is Geoffrey Vickers, "Practice and Research in Managing Human Systems—Four Problems of Relationship," *Policy Sciences* (February 1978), Vol. 9, No. 1, 1–8. Sir Geoffrey Vickers has made a major contribution to human systems knowledge through a life of writing and practice spanning over nine decades. The reader is urged to become familiar with his writings, beginning with his *The Art of Judgment: A Study of Policy Making* (New York: Basic Books, Inc., 1965), to more fully understand the role of values, appreciation set, and extrarational processes in the policymaking process. One of the earliest, and now classic, statements on the requirement for social science knowledge to serve policymaking is Robert S. Lynd, *Knowledge for What?* (Princeton, NJ: Princeton University Press, 1948). A more recent collection of papers dedicated to the same subject is Stuart S. Nagel, ed., *Policy Studies and the Social Sciences* (Lexington, MA: D. C. Heath and Company, 1975). In addition to the Thomas S. Kuhn reference 5 his *The Essential Tension: Selected Studies in Scientific Tradition and Change* (Chicago: University of Chicago Press, 1978) amplifies on his classic 1962 study of the nature of scientific knowledge evolution and revolution. For a theory of knowledge that is consistent with Michael Polyani's theory of personal knowledge and the theory of systems analysis presented here see Kenneth E. Boulding, *The Image: Knowledge in Life and Society* (Ann Arbor, MI: University of Michigan Press, 1956). Kenneth Boulding has played an important role in the development of General Systems theory. *The Meaning of the 20th Century: The Great Transition* (New York: Harper and Row, 1964). For a study toward a higher sense of purpose in policy design of human systems see Erich Jantsch, *Design For Evolution: Self Organization and Planning in the Life of Human Systems* (New York: George Braziller, 1975). This is the only one of many relevant works in the Library of Systems Theory and Philosophy Series edited by Ervin Laszlo. On the relationship of knowledge and values change to a new paradigm for human evolution read Jonas Salk, *The Survival of the Wisest* (New York: Harper and Row, 1973). All available works of Michael Polyani are relevant to this chapter. For 19th century conceptions of the scientific method see Ronald N. Giere and Richard S. Westfall, *Foundations of Scientific Method: The Nineteenth Century* (Bloomington, IN: Indiana University Press, 1973). Ernest Nagel, *The Structure of Science: Problems in Logic of Scientific Explanation* (New York: Harcourt, Brace and World, Inc., 1961), and Abraham Kaplan, *The Conduct of Inquiry: Methodology for Behavioral Research* (San Francisco: Chandler Publishing Co., 1964) are both useful. For an earlier view of the scientific method for systems analysis see Russell L. Ackoff, *Scientific Method: Optimizing Applied Research* (New York: John Wiley, 1962). The historical epistemological positions and debates of John Locke, George Berkeley, Alfred North Whitehead, Bertrand Russell, John Dewey, and Arthur Bentley; and even earlier, Sir Francis Bacon (1560–1626) as the father of inductive reasoning and analysis; and Aristotle (384–322 B.C.) for his deductive analytical thinking have impacted on the development of this chapter and the Chapter 6 research categories. See, for instance, Sir Francis Bacon's *The Advancement of Learning,* William Aldis Wright, Ed., 2nd ed. (Oxford: Oxford University Press, 1880) for an inductive analysis of learning and knowledge to the late 16th Century; and J. E. C. Welldon, translator, *The Nicomachean Ethics of Aristotle* (London: MacMillan and Co., Limited, 1908), original text 1881. In Book VI, Chapter III Aristotle lists the five means by which the soul arrives at truth as: art, science, prudence, wisdom, and intuitive reason. The relationship of scientific knowledge to policymaking has been a focus of attention since the days of Aristotle and is a fundamental concern of the policy sciences. For a recent National Academy of Sciences report on the subject see *Knowledge and Policy: The Uncertain Connection* (Washington, DC: National Academy of Science, 1978). And for the role of policy scientists in the expansion of knowledge see Harold D. Lasswell, *A Pre-View of Policy Sciences* (New York: American Elsevier, 1971), Chapter 8.

CHAPTER 2, THEORY AND MODELS

Chapter Notes and References

6. My discussion of theory owes an intellectual debt to Karl Deutsch, the distinguished political scientist. In particular his Presidential Address to the 1970 meeting of the American Political Science Association (APSA) in New York on September 10, 1970 entitled, "Political Theory and Political Action."

7. William Gray, in his "In Memorium" to Ludwig von Bertalanffy introducing the December 1972 *Academy of Management Journal,* Volume 15, No. 4, p. 404 stated the case for continual evolution of theory extremely well with the words: "A true theory of management science must be composed of streams of ideas, concepts, and observations that maintain their form only as long as the streaming flow continues. They must not degenerate into permanently fixed structures, for then they would only be a house without inhabitants."

Bibliographic Essay

This chapter addresses only the role of theory for specific disciplines of knowledge. See the *International Encyclopaedia* of the Social Sciences for general and discipline-oriented discussions of theory. The annual *Yearbook of the Society for General Systems Research,* Anatol Rapoport, Ed. (Ludwig von Bertalanffy and Anatol Rapoport, Eds. 1955 through 1973) is the best source for general systems theory and bibliographies. Also the annual *Proceedings* of the meetings of the Society for General Systems Research; particularly that of 1976 titled, "General Systems Theorizing: An Assessment and Prospects for the Future." For organizational theory, systems theory and systems analysis theory see the Bibliographic Essay sections for Chapters 3 and 4.

For the best writing on the application of normative models to the policymaking function see Yehezkel Dror, *Public Policymaking Reexamined* (San Francisco: Chandler Publishing Co., 1968), particularly Part IV. His Bibliographic Essay to Chapters 12-14 is a valuable source of literature on policymaking models up to 1968. For decision models see John P. Van Gigch, *Applied General Systems Theory* (New York: Harper and Row, 1974), Chapter 9 and references. For a view on the disparities between the theory and the computer simulation modeling of societal systems see Kenyon B. DeGreene, "Force Fields and Emergent Phenomena in Sociotechnical Macrosystems: Theories and Models," *Behavioral Science,* Vol. 23, 1978, pp. 1-14. For conceptual models see Kenneth E. Boulding, *The Image: Knowledge in Life and Society* (Ann Arbor, MI: University of Michigan Press, 1956). For policy models see Martin Greenberger, Matthew A. Crenson, and Brian L. Crissey, *Models in the Policy Process* (New York: Russell Sage Foundation, 1976).

For the other volumes in the Wiley Series on Systems Engineering and Analysis, edited by Harold Chestnut, see the list at the beginning of this book and *Books in Print* for those that will be published subsequent to this book.

CHAPTER 3, SYSTEMS CONCEPTS

Chapter Notes and References

8. C. West Churchman, *The Systems Approach* (New York: Dell Publishing Co., Inc., 1968), p. 232.

9. "Reality 21" is not a completely facetious offering. The concept itself does not predetermine answers to the centralization versus decentralization and freedom versus order issues so well stated by E. F. Schumacher in his *Small is Beautiful: Economics as if People Mattered* (New York: Harper and Row, 1973).

10. Robert S. Lynd. *Knowledge for What? The Place of Social Science in American Culture* (New York: Grove Press, Inc. 1964), p. 12. First published in 1939.

11. Ervin Laszlo. *The Systems View of the World* (New York: George Braziller, 1972).

Bibliographic Essay

One of the earlier, and still valuable, collections of papers on the system approach to the analysis of living phenomena was F. E. Emery, Ed., *Systems Thinking* (Middlesex, England: Penguin Books, Ltd., 1969). At about the same time period C. West Churchman at Berkeley was writing his two early systems books: *The Systems Approach* (New York: Dell Publishing Co., Inc., 1968) and *Challenge to Reason* (NewYork: McGraw-Hill Book Co., 1968). In Ervin Laszlo's *The Systems View of the World: The Natural Philosophy of the New Developments in the Sciences* (New York: George Braziller, 1972) is an excellent bibliography (pp. 121–126) subtitled: "A concise guide to recent books which articulate, in relatively non-technical language, the interdisciplinary foundations of the contemporary systems view." In the same year Laszlo published his more technical, *Introduction to Systems Philosophy* (New York and London: Gordon and Breach, 1972) and organized "The International Library of Systems Theory and Philosophy" under his editorship. "The Wiley Series on Systems Engineering and Analysis," under the editorship of Harold Chestnut has contributed to the development of systems concepts over a 15-year period. This book is the most recent in that series. As part of that series Ralph F. Miles, Jr., Ed., *Systems Concepts: Lectures on Contemporary Approaches to Systems* (New York: John Wiley & Sons, 1973) published a California Institute of Technology lecture series entitled: "Systems Concepts for the Private and Public Sectors," which is particularly relevant. The application of systems concepts to urban and environmental problems was stressed in the Fifteenth Annual Meeting of the Society for General Systems Research in Boston in 1969. The selection of papers from that meeting is in Milton D. Rubin,

Ed., *Systems in Society* (Washington, DC: Society for General Systems Research, 1973). Alice M. Rivlin in *Systematic Thinking for Social Action* (Washington, DC: The Brookings Institution, 1971) published her "H. Rowan Gaither Lectures in Systems Science" which addressed the problem of applying systems concepts—specifically macro cost-benefit analyses—to alternative governmental social programs. Roger A. Kaufman applied systems concepts to education in *Educational System Planning* (Englewood Cliffs, NJ: Prentice-Hall, Inc., 1972). For the systems approach as applied to a long list of subjects from water management to health care see the current Books in Print (New York: R. R. Bowker Company). For a humorously written fatalistic view of the inability of systems to function in spite of, or because of, applied systems concepts, read John Gall, *Systemantics: How Systems Work and Especially How They Fail* (New York: Quadrangle/The New York Times Book Co., 1975).

CHAPTER 4, SYSTEMS ANALYSIS

Chapter Notes and References

12. Eric Berne. *Intuition and Ego States: The Origins of Transactional Analysis* (San Francisco: Harper and Row, 1977), Editor's Preface by Paul McCormick, p. x.

13. See E. S. Quade and W. I. Boucher, eds., *System Analysis and Policy Planning: Applications in Defense* (New York: American Elsevier Publishing Co. Inc., 1968), p. 3.

14. Ibid., p. 7.

15. Congressional committees reviewed the entire Planning, Programming, and Budgeting system in a series of hearings to two subcommittees. See the three volumes of *The Analysis and Evaluation of Public Expenditures: The PPB System, A Compendium of Papers submitted to the Subcommittee on Economy in Government of the Joint Economic Committee of the Congress of the United States. Vol I, II, and III, 91st Congress, 1st Session,* (Washington, DC: U.S. Government Printing Office, 1970). Henry S. Rowen, the President of RAND Corporation from 1967 to 1972 and Assistant Director of the Bureau of the Budget in the Executive Office of the President from 1965 to 1967 answered the question of "How Has It (PPBS) Worked?" with the statement: "On the whole, not well. The obstacles to satisfactory decision making I described earlier have impeded the successful use of PPBS." See Henry S. Rowen "Planning-Programming-Budgeting Systems," in *Systems Concepts: Lectures on Contemporary Approaches to Systems,* (California Institute of Technology, 1973), p. 171.

16. Since Paul Dickson described the 1971 status of "cerebral supermarkets" in his *Think Tanks* (New York: Atheneum, 1971) the phenomenon of institutions dedicated to policy analysis and systems analysis has continued to expand on university campuses, within local, state, and national governments, in private industry and in developed and developing nations of the world.

17. See Chapter 5 for Yehezkel Dror's contribution to policy sciences theory and practice.

Bibliographic Essay

For the branch of systems analysis theory that emerged from academic research see (1) the works of Ludwig von Bertalanffy, Walter Buckley, J. A. Miller, O. R. Young, Kenneth Boulding, and Anatol Rapaport for General Systems Theory; (2) Norbert Wiener and W. Ross Ashby for cybernetic systems; (3) Karl W. Deutsch, Colin Cherry, and C. E. Shannon for Communications Systems; (4) Talcott Parsons and Amitai Etzioni for Social Systems; (5) Morton Kaplan, Stanley Hoffman, George Modelski, Charles McClelland, Richard Rosecrance, and Sidney Verba for International Systems; (6) Gabriel Almond, Robert Dahl, Karl Deutsch, and David Easton for Political Systems; and (7) the human factors literature for the psychologically based person-system linkages. The *Annals* and *Proceedings* of the Society for General Systems Research have become an excellent source for cross-discipline papers from the systems viewpoint.

For the branch of systems analysis originating with weapons systems studies during World War II and spurred by the economically based analysis techniques developed at the RAND Corporation and implemented throughout the U.S. Department of Defense under Secretary of Defense Robert S. McNamara, see the first book edited by E. S. Quade, *Analysis for Military Decisions* (Chicago: Rand McNally, 1964). Claims for the use of systems analysis predate World War II, however, For instance, Mr. Shim Moon-Taik, President of the Military Operations Research Society of Korea announced to a Pacific Conference on Operations Research in Seoul, in April 1979, that operations research techniques were being applied to military civil and business problems as early as 1394, at the beginning of the Yi Dynasty, in Korea. The model in existence then was called "Kyung Kook Dae Chun" and included government organization, rules for social life, and plans for industrial complexes. The model was successfully used, according to Mr. Shim for 500 years (see the *Korea Herald,* April 24, 1979, p. 3.). The literature of systems analysis definitely has a Western bias which hopefully will be overcome in the 1980s.

Charles J. Hitch's *Decision-Making for Defense* (Berkeley, CA: University of California Press, 1965) is considered the first authoritative inside discussion of the Planning-Programming-Budgeting system and the cost-effectiveness techniques that Dr. Hitch helped Secretary of Defense McNamara install in Washington beginning in 1961. The Planning-Programming-Budgeting System (PPBS) literature has proliferated since that time. The reader interested in the 1960s debates over systems analysis and the PPBS subject should refer to the rich source of literature in the three-volume *The Analysis and Evaluation of Public Expenditures: The PPB System, A compendium of papers submitted to the Subcommittee on Economy in Government of the Joint Economic Committee, Congress of the United States, 91st Congress, 1st Session* (Washington, DC: U.S. Government Printing Office, 1969). Part of this same debate is also found in *Planning, Programming, Budgeting,* Inquiry of the Subcommittee on National Security and International

Operations, Senator Henry M. Jackson, Chairman for the Committee on Government Operations, U.S. Senate (Washington, DC: U.S. Government Printing Office, 1970). During the same period the U.S. General Accounting Office distributed its: *Glossary for Systems Analysis and Planning-Programming-Budgeting* (October 1969) which clearly shows reliance of systems analysis to that date as being on cost-benefit and cost-effectiveness economic and mathematical formulas with stated recognition of the need for judgment but no qualitative analytical tools to assist in that judgment. PPBS has continued, in varying interpretations, to be applied by Departments and Agencies of the U.S. Government through the 1970s. Zero Base Budgeting (ZBB) was an add-on concept under the Administration of President Jimmy Carter in January 1978. PPBS was designed in an era of increasing DOD budgets and ZBB was designed as a control device to avoid further increases not thoroughly justified. Both systems fall into the economically rational model discussed in Chapter 5 and create new demands for systems analysis. For ZBB see Austin L. Allan, *Zero-Base Budgeting: Organizational Impact and Effects* (New York: AMACOM, 1977).

E. S. Quade's second book on systems analysis with W. I. Boucher: *Systems Analysis and Policy Planning* (New York: American Elsevier, 1968) describes and demonstrates many of the basic techniques and his next volume: *Analysis for Public Decisions* (New York: American Elsevier, 1975) recognizes the place of political realities, judgment, and intuition in formulating public policy—one of the major thrusts of this volume. For systems analysis applied to computer sciences see (1) Philip C. Semprevino, *Systems Analysis: Definition, Process and Design* (Palo Alto, CA: Science Research Associates, Inc., 1976); (2) Paul Gross and Robert D. Smith, *Systems Analysis and Design for Management* (New York: Dun-Donnelley Publishing Corporation, 1976)—the former is business-oriented and the latter covers public and private computer-based information systems and includes a useful list of journals and magazines; and (3) Robert J. Thierauf, *Systems Analysis and Design of Real-Time Management Information Systems* (Prentice-Hall, 1975).

The current Books in Print section on systems analysis should be consulted and provides a reality test, through comparison with previous years' editions, for my thesis that systems applications and literature are increasing.

CHAPTER 5, POLICY SCIENCES AND QUALITATIVE TOOLS

Chapter Notes and References

18. The plural form "Policy Sciences" as used grammatically in the singular seems to have become accepted now. Dror explains this contradiction as the desire "to emphasize the multiple components on one hand, the basic unity on the other hand." *Policy Sciences,* Vol. 1, No. 1 (Spring 1970), p. 137.

19. Harold D. Lasswell, "The Policy Orientation," in *Policy Sciences* Daniel

Lerner and Harold D. Lasswell, eds., (Stanford, CA: Stanford University Press, 1951), Chapter I. This seminal article is considered the beginning of Policy Sciences as a new interdiscipline.

20. Harold D. Lasswell. *A Pre-View of Policy Sciences* (New York: American Elsevier, 1971).

21. *Policy Studies Journal* (Urbana, IL. The Policy Studies Organization, University of Illinois) first edition was Autumn 1972. Since the emergence of *Policy Sciences* (1970) and the *Policy Studies Journal* (1972) a number of other professional journals in the policy area have been initiated. See the bibliographic essay, following.

22. Dickson, *Think Tanks,* op. cit.

23. The policy scientist agrees with Isaiah Berlin in his defense of free will over determinism. See his *Historical Inevitability* (London: Oxford University Press, 1954). Geoffrey Vickers also describes the view well in the final chapter "The End of Free Fall" in *Value Systems and the Social Process* (New York: Basic Books, 1968).

24. The term "policy analysis" is used in the literature in two senses. First is policy analysis as a normative research aid for identification of preferable policy alternatives. Secondly, in behavioral research to describe the context, content, and origin of discrete policies and policymaking systems. Both senses are useful in systems analysis.

25. E. Joseph Piel and John G. Truxal in *Technology: Handle With Care* (New York: McGraw Hill, 1975), pp. 84–95 provide a supporting statistical explanation for this by hypothesizing that public indifference to life and property loss on highways, and elsewhere, occurs as long as the risk of death (e.g., in auto travel) remains about the same as the risk of death for normal living. They cite Chauncey Starr, "Social Benefit Versus Technological Risk," *Science* (Sept. 19, 1969), p. 1232. In the case of the United States Vietnam War reaction there were extra-rational variables other than implicit awareness of average death and injury statistics. The Japanese government has made traffic fatalities an issue through strict implementation of safety standards and operating procedures resulting in significantly reduced traffic accidents and fatalities in the 1970s.

26. See Table 1.2 and the Michael Polyani works in reference 2.

27. Alfred North Whitehead, *Process and Reality* (New York, 1929).

28. Eric Berne in Intuition and Ego States: *The Origins of Transactional Analysis,* Paul McCormick, ed., (San Francisco: Harper and Row, 1977) defines intuition as ". . . unverbalized processes and observations based on previously formulated knowledge which has become integrated with the personality through long usage and therefore functions below the level of consciousness . . ." p. 1. The observation of isomorphy between portions of Polyani's theory of personal knowledge and Berne's concept of intuition is mine. Berne makes no reference to Polyani's earlier work in this book. Polyani notes that tacit knowing is manifested most clearly in the act of understanding or comprehending disjointed parts into a comprehensive whole and that as a process this has been carefully traced by the psy-

chology of Gestalt but that psychologists describe our perception of Gestalt as a passive experience whereas Polyani describes tacit knowledge as being shaped by the knower's personal action. See Polyani's, *The Study of Man,* op. cit., p. 28.

29. See, for instance, Eleanor Farrar McGown, "Rational Fantasies," *Policy Sciences,* Vol. 7 (1976), pp. 439-454. Also George W. Ball's essay on the Pentagon Papers titled "The Trap of Rationality," which ended with the conclusion: "Thus, in the long run, what misled a group of able and dedicated men was that, in de-personalizing the war and treating it too much as an exercise in the deployment of resources, we ignored the one supreme advantage possessed by the other side: the non-material element of will, of purpose and patience, of cruel but relentless commitment to a single objective regardless of human cost—something that, occurring on our side, we would call 'exceptional patriotism' but which, when displayed by the enemy, necessarily appeared as 'irrationality' or "fanaticism" a challenge to faith in quantified resources. Yet that was the secret of North Viet-namese success—rebuke of spirit to the logic of number." *Newsweek,* July 26, 1971, p. 64.

30. Robert M. Krone, "Policy Sciences and Civil-Military Systems," *Journal of Political-Military Sociology,* Vol. 3, No. 1 (Spring 1975), pp. 81-84.

31. Herman Kahn and Irwin Mann, "Ten Common Pitfalls," *RAND Research Memorandum RM 1937,* July 17, 1957. The authors use the "Butch" meaning a "completely mistaken technical notion or fact," but the concept is equally appli-cable to a mistaken cultural assumption or the neglect of a critical cultural variable.

32. Geoffrey Vickers, *Value Systems and Social Process* (New York: Basic Books, 1968), final chapter.

Bibliographic Essay

The basic theoretical reading for anyone interested in Policy Sciences are the works of Harold Lasswell cited in the references plus his entry "Policy Sciences" in the *International Encyclopedia of Social Sciences,* (New York: The Macmillan Company of the Free Press, 1968), Vol. 12, pp. 181-189; and his "The Emerging Concepts of Policy Sciences," *Policy Sciences,* Vol. 1, No. 1 (Spring 1970). It is interesting to note that all of Harold Lasswell's references in his *International Encyclopedia of the Social Sciences* description of Policy Sciences are to his own works and Dror's 1968 book is the only reference Lasswell uses in his 1970 *Policy Sciences* article. This is an indication of the youth of the interdiscipline of Policy Sciences which dates back to Harold Lasswell's 1951 article previously cited but which made little progress until Dror's works of 1968 and 1971. Yehezkel Dror's three Policy Sciences books which converted Harold Lasswell's concepts into a solid theoretical base were: *Public Policymaking Reexamined* (San Francisco: Chandler Publishing Co., 1968); *Design for Policy Sciences* (New York: American Elsevier Publishing Co., Inc., 1971); and *Ventures in Policy Sciences* (New York: American Elsevier Publishing Company, 1971). For an introductory essay on mili-tary linkages see Robert Krone, "Policy Sciences and Civil Military Systems,"

Journal of Political and Military Sociology Vol. 3, No. 1 (Spring 1975), 71–84 and "NATO Nuclear Policy-Making," in John P. Lovell and Philip S. Kronenberg, *New Civil-Military Relations: The Agonies of Adjustment to Post-Vietnam Realities* (New Brunswick, NJ: Transaction Books, distributed by E. P. Dutton and Co., 1974), pp. 193–228.

The relevant professional journals—a 1970s addition to the literature—are *Policy Sciences: An International Journal Devoted to the Improvement of Policy Making* (began 1970) published bi-monthly by Elsevier Scientific Publishing Co., Amsterdam, The Netherlands; *Policy Analysis* (began 1975) published quarterly by the Graduate School of Public Policy at the University of California, Berkeley; the *Policy Studies Review Annual* (began 1978) Howard E. Freeman, Ed., Institute for Social Science Research, University of California, Los Angeles; *Policy Review* (began 1978) Quarterly Journal of the Heritage Foundation; *Public Policy* (began 1969) John F. Kennedy School of Government, Harvard University; *Policy and Politics* (began 1972) Sage Publishers, Ltd., London; the *Policy Studies Journal* (began 1972) published quarterly by the Policy Studies Organization, University of Illinois and the *Policy Review Annual* (began 1977) Sage Publications, Inc. A particularly valuable 1978 publication of the Policy Studies Organization was the *Policy Research Centers Directory* co-edited by Stuart Nagel and Marian Neef. This volume includes a bibliography of policy research center directories previously published and is designed to describe university and nonuniversity centers, institutes, or organizations that conduct policy studies research. The Directory contains results of responses from 107 surveyed policy research centers.

At this point in the development of policy sciences (1980) a complete bibliography of works contributing to the idea of policy sciences, as described and defined in Chapter 5, would fill a volume in itself. For the best bibliographies available up to 1971 see the three basic works of Yehezkel Dror previously cited. A more recent bibliography on public policy is Douglas Ashford et. al. *Comparative Public Policy: a Cross-National Bibliography* (Beverly Hills, Calif: Sage Publications, 1978). In addition to my own policy sciences works cited elsewhere see Robert Krone, "The Role of Policy Sciences in the Future Development of the Republic of Korea," *Korea Observer* (Seoul: The Academy of Korean Studies) Vol. X, No. 3 (Autumn 1979), pp. 298–309.

The best sources for future studies are the publications of the World Future Society—particularly, *The Futurist: A Journal of Forecasts, Trends and Ideas about the Future,* (began in 1967), published bi-monthly; and the *World Future Society Bulletin,* published bi-monthly. A review of past editions of these publications will lead the reader to most relevant future literature. A particularly good recent work is: Christopher Freeman and Marie Jahoda, Eds., *World Futures: The Great Debate* (New York: Universe Books, 1978). In this volume, pages 393–408, there is an excellent futures bibliography.

As policymaking is the process of converting values into action to achieve goals the entire subject of values and their relationship to the phenomenon of choice is relevant to Policy Sciences. Unfortunately there is as yet no single bibliographic source for values in the policy process. The fields of sociology, anthropology, aesthetics, psychology, theology, and epistemology all have linkages to values. For instance Clyde Kluchkhohn, "The Scientific Study of Values and Contemporary Civilization," *Zygon,* Vol. 1 (September 1966), 230-243 is excellent from the sociological viewpoint. The Hudson Institute, under the direction of Herman Kahn, has done considerable investigation into values and policy much of which has been published in the works of Kahn and his associates. Sir Geoffrey Vickers in his works previously cited has made significant contributions in the relationship of values to social policy. His *Value Systems and Social Process* (New York: Basic Books, 1968) and *Freedom in a Rocking Boat: Changing Values in an Unstable Society* (Middlesex, England, Penguin Books, 1970) are classics. Sir Geoffrey explores the linkages between the arts, the humanities, and the sciences in his "Rationality and Intuition," in Judith Wechsler, Ed., *On Aesthetics in Science* (Cambridge, MA: The MIT Press, 1978). For a Finnish study in policy sciences see J. P. Roos, *Welfare Theory and Social Policy: A Study in Policy Science* (Helsinki: Societas Scientiarum Fennica, 1973). For the concept of political feasibility see Dror, *Design for Policy Sciences,* Chapter 9, and also his *Ventures in Policy Sciences,* Chapter 8.

CHAPTER 6, RESEARCH METHODOLOGY
FOR SYSTEMS ANALYSIS

Chapter Notes and References

33. For a more detailed theoretical discussion of this definition see Gordon K. C. Chen, "An Anatomy of Problem Solving Systems," in Milton D. Rubin, Ed., *Systems in Society* (Washington, DC: Society for General Systems Research, 1973), pp. 17-29. See also "Problem Solving" by Donald W. Taylor in the *International Encyclopedia of The Social Sciences,* David L. Sills, Ed., (New York: The MacMillan Co. and The Free Press, 1968), Vol. 12, pp. 504-511.

34. Richard E. Truman, in his *An Introduction to Quantitative Methods for Decisionmaking,* 2nd ed. (New York: Holt Rinehart and Winston, 1977) after 650 pages of presentation of methods provides his "Reflections on Decision Analysis and its Future Prospects," as follows: "Although many effective and important decision analysis studies have been performed over the years, there is a strong feeling, among both professionals in the field and business executives, that decision analysis has not made the contribution to business that it could have. One of the major reasons for this appears to be the communications gap between decision analysts and management."

35. The form of the question comes from the classic study of politics by Harold D. Lasswell, *Politics: Who Gets What, When, How?* (New York: Whittlesey House, 1936: Reprinted New York: Peter Smith, 1950).

36. Note that this idea utilizes rather than bypasses the representative structure

of democratic politics. For the Delphi method of surveying expert opinion see Harold A. Linstone and Murray Turoff, Eds., *The Delphi Method: Techniques and Applications* (Reading, MA: Addison-Wesley Publishing Co., 1975). For a variant of this idea using teaching computer technology to exchange information and opinions between experts and a cross-section of the public on long-range planning issues see Stuart Empleby, "Citizen Sampling Simulations: A Method for Involving the Public in Social Planning," *Policy Sciences*, Vol. 1, No. 3 (Fall 1970), pp. 361–375. Some economic, technological, and political feasibility and desirability considerations are also provided. The Center for Policy Research in New York City and Washington, DC developed in theory a complete citizens-leaders interactive system called "Minerva" (for "Multiple Input Network for Evaluating Reactions, Votes and Attitudes"). Minerva was the Roman Goddess of political wisdom. Eugene Leonard, et. al., "Minerva: A Participatory Technology System," undated *Center for Policy Research, Inc.,* monograph given the author at the New York offices July 1974.

Bibliographic Essay

For scientific explanation and procedure see Ernest Nagel, *The Structure of Science: Problems in the Logic of Scientific Explanation* (New York: Harcourt, Brace and World, 1961) and Abraham Kaplan, *The Conduct of Inquiry* (San Francisco: Chandler, 1964). The methodology of *Systems Analysis and Policy Sciences* combines the functional-structural methods of the social sciences, the economic and engineering based systems analysis and management sciences, and the normative models of policy sciences. See *Functionalism in the Social Sciences: The Strength and Limits of Functionalism in Anthropology, Economics, Political Science, and Sociology,* edited by Don Martindale, Monograph 5 in a series sponsored by the American Academy of Political and Social Science, Philadelphia, February 1965. Also Donald P. Eckman, Ed., *Systems Research and Design* (New York: John Wiley, 1961) and Arthur D. Hall, *A Methodology for Systems Engineering* (Princeton, NJ: Van Nostrand, 1962). For theoretical behavioral models read Albert Einstein's works. For instrumental-normative models see Dror, *Public Policymaking Reexamined,* p. 22. For two good collections of papers on methodologies for policy see Stuart S. Nagel, *Policy Studies and the Social Sciences* (Lexington, MA: Lexington Books, 1975); and Frank P. Sciolo, Jr. and Thomas J. Cook, Eds., *Methodologies for Analyzing Public Policies* (Lexington, MA: Lexington Books, 1975).

    The measuring of values is partially covered in Kenneth J. Arrow, *Social Choice and Individual Values,* 2nd ed. (New York: John Wiley, 1963). Some futures research methodologies can be found in Harman W. Willis, *An Incomplete Guide to the Future* (San Francisco: San Francisco Book Co., Inc., 1976). A journal covering methods used by major research and development corporations in the United States is *Research Management* (began 1958), Industrial Research Institute, New York.

CHAPTER 7, EVALUATION AND MEASUREMENT

Chapter Notes and References

37. There has been too little research in this fertile area. Much of what has been done is by psychologists and political scientists concerning political leadership. Harold D. Lasswell, again, is credited with stimulating this new field of research with his *Power and Personality* (New York: Viking Press, 1962). One of the most publicized research efforts to date was that of Yale Professor James D. Barber who did an analysis of the character-style combination of U.S. Presidents in the 20th century, ending with Richard Nixon. In September 1969 he presented a paper titled "The President and His Friends," (later to be expanded into a book) at the 65th Annual Meeting of the American Political Science Association. One of his conclusions was that "The danger is that Nixon will commit himself irrevocably to some disastrous course of action." (p. 46). The coverup of the illegal break-in of the Watergate Building Democratic Offices in Washington occurred in 1972 and led to Richard Nixon's resignation effective January 20, 1973. There continues a need for increased research into the relationship of extrarational processes and leadership—particularly one-person-centered decision systems. See also reference 29.

Bibliographic Essay

My two-step evaluation method is an adaptation of the evaluation approach to public policymaking in Dror, *Public Policymaking Reexamined*, Chapters 3 to 6. For social secondary criteria see Bertram M. Gross, *The State of the Nation: Social-System Accounting* (London: Tavistock, 1966); B. M. Russett et al., *World Handbook of Political and Social Indicators* (New Haven: Yale University Press, 1964); Eleanor Bernert Sheldon and Wilbert E. Moore, Eds., *Indicators of Social Change* (New York: Russell Sage Foundation, 1968); The *United Nations Statistical Yearbooks;* and the World Bank, *World Development Report* commencing with the 1978 edition (Oxford University Press). For evaluation of U.S. Federal policy in social agencies see Walter Williams, *Social Policy Research and Analysis* (New York: Elsevier, 1971).

As is the case with values analysis there is no consolidated bibliography of systems analysis or policy sciences related evaluation works. My forthcoming book, *Systems Evaluation,* will expand on the evaluation theory presented here and attempt to fill the void. Evaluation literature is best developed in the field of education. See, for example, Daniel I. Stufflebeam et al., *Educational Evaluation and Decisionmaking,* Phi Delta Kappa National Study Committee on Evaluation (Itasca, IL: F. E. Peacock Publishers, Inc., 1971); Scarvia B. Anderson et al., *Encyclopedia of Educational Evaluation: Concepts and Techniques for Evaluating Education and Training Programs* (San Francisco: Jossey-Boss Publishers, 1975); and Stephen Isaac, *Handbook in Research and Evaluation for Education and the Behavioral Sciences* (San Diego: Robert R. Knopp Publisher, 1971). The definition of evaluation, provided in the study of the Phi Delta Kappa National Study Committee on Evaluation, is consistent with mine and reads: "Evaluation is the process of delineating, obtaining, and providing useful information for judging decision alternatives."

CHAPTER 8, QUANTITATIVE TOOLS FOR SYSTEMS ANALYSIS

Chapter Notes and References

38. The concept of outcome arrays is developed by David I. Cleland and William R. King in *Systems Analysis and Project Management,* 2nd ed. (New York: McGraw Hill Book Co., 1975), pp. 75–111.

39. See C. West Churchman, Leonard Auerback, and Simach Sadan. *Thinking for Decisions: Deductive Quantitative Methods* (Palo Alto, CA: Science Research Associates, Inc., 1975), particularly Chapters 1 and 2.

40. E. S. Quade and W. I. Boucher, eds., *Systems Analysis and Policy Planning: Applications in Defense* (New York: American Elsevier Publishing Co., Inc., 1968), p. 78.

41. This is the same distinction made in the University of Southern California Master of Science in Systems Management degree program curriculum. It has the advantage of dealing with problems of certainty in one graduate course and with problems of uncertainty in another.

42. See Churchman, 1975, p. 16, op. cit., for a discussion of types of assumptions made in quantitative analysis.

43. Another definition of risk in mathematics is "the worst that can happen under given conditions." See John Von Neumann and Oskar Morgenstern *Theory of Games and Economic Behavior,* Science Edition (New York: John Wiley and Sons, 1964), p. 163 original copyright 1944 by Princeton University Press.

44. See E. S. Quade and W. I. Boucher, eds., *Systems Analysis and Policy Planning* (New York: American Elsevier, 1971).

45. *Ibid.,* Chapter 8 and all of Gene H. Fisher *Cost Considerations in Systems Analysis* (New York: American Elsevier, 1971).

Bibliographic Essay

As much of systems analysis developed under the assumption that a mathematical or economic model must be the central component of analysis, the quantitative variables literature is much more sophisticated than the qualitative tools literature as we move into the 1980s. A large number of journals play important roles. *Management Science* (began 1954) of the Institute of Management Sciences at Providence, Rhode Island is oriented toward managers who want to apply mathematical and scientific methods to decision-making. *Omega: International Journal of*

Management Sciences (began 1973) does the same with an international orientation. The *Operational Research Quarterly* (began 1950) is the official organ of the
Operational Research Society and is a highly technical and mathematically oriented
research journal. *Operations Research* (began 1952) is the journal of the Operations
Research Society of America, Baltimore, MD. Related to the last two journals listed
are: *Interfaces; OR/MS Today;* and *Mathematics of Operations Research* published
under the sponsorship of The Institute of Management Sciences (TIMS) and the
Operations Research Society of America. *Decision Sciences* is the journal of the
American Institute for Decision Sciences and is designed for teachers of business
administration. The *International Journal of Systems Science,* London, covers
theory and practice of mathematical modeling, simulation, optimization, and
control for economic, industrial, and transportation systems. The *Logistics Spectrum* of the Society for Logistics Engineers publishes technical papers across the
logistics field. The *Quality Engineer* is the journal of the Institute of Quality
Assurance oriented toward technical, statistical, and metrology articles. *Science*
is a standard for both quantitative and qualitative issues and research. The *Management Review* of the American Management Association provides an excellent mix
of quantitative and qualitative research results as well. The *Journal of International
Business Studies* a joint publication of the Academy of International Business
and Rutgers, The State University of New Jersey, Graduate School of Business
Administration is oriented toward international financial management and multinational corporation research. With the increasing applications of small decen-
tralized computer systems *Mini-Micro Systems* (Denver: Cahners Publishing Co.)
allows the reader to keep up with that rapidly changing field in computer sciences.
Datamation (Barrington, IL: Technical Publishing Co.) covers the entire spectrum
of data processing technology.

For quantitative approaches, models and tools textbook sources refer to Richard
I. Levin and Charles A. Kirkpatric, *Quantitative Approaches to Management,* 4th
ed., (New York: McGraw-Hill, 1978) with accompanying Instructors Manual;
Harold Bierman, Jr., Charles P. Bonini, and Warren H. Hausman, *Quantitative
Analysis for Business Decisions,* 5th ed. (Homewood, IL, Richard D. Irwin, Inc.,
1977) with accompanying Solutions Manual; C. West Churchman, Leonard Auer-
bach, and Simcha Sadan, *Thinking for Decisions: Deductive Quantitative Methods*
(Palo Alto, CA: Science Research Associates, Inc., 1975) with accompanying
Solutions Manual; Paul Jademus and Robert Frame, *Business Decision Theory*
(New York: McGraw-Hill, 1969) with Instructors Manual; William A. Spurr and
Charles P. Bonini, *Statistical Analysis for Business Decisions,* rev. ed. (Homewood,
IL: Richard D. Irwin Inc., 1973); Ya-lun Chou, *Probability and Statistics for
Decision Making* (New York: Holt, Rinehart and Winston, 1972).

For computer applications see: Davis B. Bobrow and Judah L. Schwartz, *Computers and the Policy-Making Community: Applications to International Relations*
(Englewood Cliffs, NJ, Prentice-Hall, Inc., 1968); Donald H. Sanders, *Computers
in Business: An Introduction* 3rd ed., (New York: McGraw-Hill Book Co., 1975)
with Instructors Manual; Norman Sanders, *A Manager's Guide to Profitable Computers* (New York: AMACOM, 1979); George K. Chacko, *Computer-Aided*

Decision-Making (New York: American Elsevier, 1972) and his later two-volume set: *Applied Operations Research: Systems Analysis in Hierarchical Decision-Making.* Volume one is subtitled *Systems Approach to Public and Private Sector Problems* and Volume 2 is subtitled *Operations Research to Problem Formulation and Solution* (New York: North Holland, Elsevier, 1976). Jay W. Forrester of MIT is credited with the pioneering work in dynamic computer systems applications to industrial and social problems. See his *Industrial Dynamics* (Cambridge, MA: M.I.T. Press, 1961); *Urban Dynamics* (Cambridge, MA: M.I.T. Press, 1969); and his *World Dynamics* (Cambridge, MA: Wright-Allen Press, Inc., 1971). Further elaboration on the Forrester model is in Michael R. Goodman, *Study Notes in System Dynamics* (Cambridge, MA: Wright-Allen Press, Inc., 1974) and *Dynamo User's Manual* 5th Ed., by Alexander L. Pugh III (Cambridge, MA: The M.I.T. Press, 1976). The Forrester dynamics model became the basis for analysis of the Club of Rome under the leadership of Dr. Aurelio Peccei and the research of Dennis L. Meadows and colleagues toward the limits to growth and alternatives to growth international movement. See Dennis L. Meadows, et. al. *The Limits to Growth,* 2nd ed. (New York: Universe Books, 1974). The first International Limits to Growth Conference was held in Houston, Texas in 1975; the second changed its name to Alternatives to Growth in 1977. Efforts such as the creation of a new economic order also emerged from these studies and the impetus of the Club of Rome. See RIO: *Reshaping the International Order,* Jan Tinbergen, Coordinator, A Report to the Club of Rome (New York: E. P. Dutton & Co., Inc., 1976).

John von Neumann and Oskar Morgenstern are credited with developing game theory in their *Theory of Games and Economic Behavior* (Princeton, NJ: Princeton University Press, 1944). An addition to the theory was Anatol Rapoport, *Two-Person Game Theory: The Essential Ideas* (Ann Arbor, MI: The University of Michigan Press, 1966). For a simpler introduction see Morton D. Davis, *Game Theory: A Nontechnical Introduction* (New York: Basic Books, 1970).

CHAPTER 19, THE ROLE OF THE SYSTEMS ANALYST

Chapter Notes and References

46. I have expanded on this fact in Robert Krone "Systems Management and The Human Condition," *Proceedings of the Silver Anniversary International Meeting of the Society for General Systems Research* (SGSR), London, England, August 20-24, 1979.

47. A publication dedicated to psychological inner synthesis of the individual is *Synthesis: The Realization of the Self* (Psychosynthesis Institute, San Francisco). Synthesis I, 1974 and Synthesis II, 1975 were revised for publication in 1977 and 1978.

48. From Polonius' farewell advice to his son, Laertes, on Laertes' departure from Denmark to France in Shakespeare's Hamlet, Act I, Scene III, line 78.

49. Yehezkel Dror, *Crazy States* (Lexington, MA: Heath Lexington Books, 1971). In this book Dror assesses the probabilities of the emergence of crazy groups or nation states in the international system; the fallacies of American strategic studies that tend to neglect such probabilities or their impacts; and provides some "preferable countercraziness strategies" for decisionmakers to consider.

50. Major General Jasper A. Welch, Jr., Assistant Chief of Staff, Studies and Analysis, USAF, "Systems Analysis and DOD" in *Supplement to the Air Force Policy Letter for Commanders* (Office of the Secretary of the Air Force, January 1977), Washington, DC: 1–1977, p. 16.

Bibliographic Essay

In Harold D. Lasswell's twenty-year later discussion of policy sciences (his first being in 1951), *A Pre-View of Policy Sciences* (New York: American Elsevier, 1971), he sketches policy science careers (Chapter 1), professional identity of policy scientists (Chapter 7), and professional training requirements (Chapter 8). Yehezkel Dror also discusses the professional role of policy analysts in government in his Chapter 21, "Policy Analysts: A New Professional Role in Government," in *Ventures in Policy Sciences* (1971). In that discussion Dror outlines his view of the differences between systems analysts and policy analysts. On the role of policy sciences in the improvement of the policymaking capabilities of politicians, lawyers, senior civil servants, and systems analysts (or "experts" as he termed them) see his Chapter 18 of *Public Policymaking Reexamined* (1968).

The actual and potential roles of policy scientists, policy analysts, systems analysts, or consultants (those four categories are not always distinct and their definitions are sensitive to the individual study) have received considerable attention—only a representative few will be listed here. *Walter Williams, Social Policy Research and Analysis: The Experience in the Federal Social Agencies* (New York: Elsevier, 1971) from the viewpoint of structural and bureaucratic environments for the analyst. On the role of experts in government see W. A. Johr and H. W. Singer, *The Role of the Economist as Official Advisor* (London: Allen & Unwin, 1955). An early study on congressional staffs is Kenneth Kofmehl *The Professional Staff of Congress* (West Lafayette, IN: Purdue University Studies, 1962). Harold Wilensky exposed the problems of adding qualified staff to labor unions in *Intellectuals in Labor Unions* (Glencoe, IL: Free Press, 1956). See also his later work: *Organizational Intelligence: Knowledge and Policy in Government and Industry* (New York: Basic Books, 1967).

The relationship of science to government has received considerable attention over the past two decades with classic works such as C. P. Snow, *Science and Government* (New York: New American Library, 1962); and Don K. Price, *The Scientific Estate* (Cambridge, MA: Harvard University Press, 1965). Harvey Brooks, *The Government of Science* (Cambridge, MA: M.I.T. Press, 1968) and Harold Wool, *The Military Specialist* (Baltimore: Johns Hopkins Press, 1968) are two different perspectives; as is Harold D. Lasswell, "Must Science Serve Political Power?" *American Psychologist,* Vol. 25 (1970), pp. 117–123.

For a discussion of the staff increases necessary with the implementation of PPBS in the United States government to the year 1969 see "Staff Increases for PPBS," Attachment 1 to *The Analysis and Evaluation of Public Expenditures: the PPB System*. A compendium of papers submitted to the Subcommittee on Economy in Government of the Joint Economic Committee, 91st Congress, 1st Session, Congress of the United States, Vol. 2, Part IV titled "The Current Status of the Planning-Programming-Budgeting System" (Washington: U.S. Government Printing Office, 1969). At page 636 of that volume it cites a total of 825 professional analytic positions for the PPB system in 21 agencies not including the military side of the Defense Department, the Central Intelligence Agency, Small Business Administration, Civil Service Commission, or Tennessee Valley Authority. Another 115 support positions were added for those professional positions.

An investigation by the United Press International into the world of consulting in Washington, D.C. showed 17,963 consultants or advisors on the federal payroll in 1978 for approximately 34,000 contracts totaling $1.8 billion per year. Even discounting those figures for the consultant category being utilized as a label for numerous other services provided to the U.S. government, the figure is large. A useful behavioral research effort would be to attempt to identify the numbers of analysts involved in public and private systems throughout the world.

The answers to the questions of whether analysts are working for healthy or pathological systems and which ones should be terminated can come only from values and values analysis—a portion of systems analysis and policy sciences in need of considerable sophistication. Aristotle maintained that: "Every art, every science, every action or purpose, aims at some good." See J. E. C. Welldon, Translator, *The Nichomachean Ethics of Aristotle* (London: MacMillan and Co., Limited, 1908), original text 1881, Book I, Chapter I, p. ix. Modern systems analysts have an obligation to define for themselves, and for their clients, what is good in relationship to which sets of values.

CHAPTER 20, ANALYSIS AND COMPLEX PROBLEMS

Chapter Notes and References

51. See the Foreward to Donella H. Meadows et al., *The Limits to Growth: A Report for the Club of Rome's Project on the Predicament of Mankind*, 2nd ed. (New York: Universe Books, 1974).

52. Jay W. Forrester, MIT, "Counterintuitive Behavior of Social Systems," *Technology Review*, (January 1971), pp. 53-68. A related newer concept is autopoiesis—or the interaction and complementarity of the spontaneous (not planned or controlled) and the artificial (man-made) aspects of social orders. For the emerging literature on autopoiesis see the publications of the Society for General Systems Research.

53. The poetry of Alfred Lord Tennyson in his prologue to "In Memoriam" has yet to be surpassed on the point of humility. It reads:

our little systems have their day:
  They have their day and cease to be:
  They are but broken lights of Thee,
And thou, O Lord, art more than they.

54. Dr. Lewis Thomas, President of the Memorial Sloan-Kettering Cancer Center, New York in an article titled "Hubris in Science?", *Science,* Vol. 200 (June 30, 1978), pp. 1459–1462.

55. This projection has been made by Wilfred A. Kraegel in his article, "Futurizing a City: A Proposal for Milwaukee's Coming Transformation," *World Future Society Bulletin* (May–June 1978). p. 7.

56. I am comfortable with the term "systems analysis" to represent the process of investigation into the phenomenon of choice that this book has explicated. The term "policy analysis" could have also been used but would have given a connotation of neglect of the tools and techniques of systems analysis as developed in the 1940s to 1970s period rather than of melding those tools with the qualitative tools of policy sciences. See also Note 24.

57. C. West Churchman in his concluding chapter of *The Systems Approach* (New York: Dell Publishing Co., 1968).

58. Arnold Toynbee, "Human Savagery Cracks Thin Veneer," *Los Angeles Times,* September 6, 1970, Section C, p. 3.

Bibliographic Essay

Good sources for obtaining a Gestalt appreciation for the complexity of systems are the *Proceedings* of the annual meetings of the Society for General Systems Research (SGSR). As an example review the Table of Contents of the *Proceedings of the 1975 Annual North American Meeting* titled "Systems Thinking and the Quality of Life." For a discussion including mathematical representations of decompositions of complexity see John P. Van Gigch, *Applied General Systems Theory* (New York: Harper and Row, 1974), Chapter 12. For a defense of the proposition that "complex systems and processes all ultimately have to be traced back to physical law" see H. Soodak and A. Iberall "Homeokinetics: A Physical Science for Complex Systems," *Science,* Vol. 201, No. 4356 (18 August 1978), 579–582.

The best collection I have found on the roles of the specialist and the generalist in government is seventy pages of abstracts from writings over the period 620 B.C. to 1965 A.D. covering the spectrum of proverbs, fables, historical episodes, and analytical writings. This collection begins the volume of testimony to the subcommittee on National Security and International Operations, Senator Henry M. Jackson, Chairman, for the Committee on Government Operations, United States Senate, titled: *Planning: Programming: Budgeting* (Washington, DC:

U.S. Government Printing Office, 1970), pp. 33-103. The selections include Aesop's Fables, Thucydides, Machiavelli, Francis Bacon, Hans Christian Anderson, Alfred North Whitehead, Winston Churchill, and others. For recent commentary by a distinguished generalist see "Breaking Out of the Double Bind," Gregory Bateson interviewed by Daniel Goleman in *Psychology Today* (August 1978), 43-51.

An excellent collection of papers on the understanding of science as a social phenomenon will be found in Bernard Barber, Ed., *The Sociology of Science* (New York: Free Press, 1962) this collection includes the writings of Talcott Parsons, Robert K. Merton, Thomas S. Kuhn, Max Weber, Edward A. Shils, and Ernest Nagel. On the dilemma of either relying on conservative and traditional institutions for policymaking or the utilization of new validated knowledge to cope with rapid societal change see Yehezkel Dror, *Public Policymaking Reexamined* (1968), Chapter 22, "The Significance of Society's Major Alternatives."

No matter how complex our human systems and their technology become it will be up to humans to accomplish the new synthesis following the decomposition of the system into its components for analysis. That new synthesis must be after the analysis of qualitative and quantitative variables with the best tools available. Systems analysis conceived in a policy sciences framework can be an instrument to help improve the human condition and to avoid the pitfalls described by Norbert Wiener in *The Human Use of Human Beings: Cybernetics and Society* (Garden City, NY: Doubleday Anchor Books, 1954), first published in 1950. I have amplified on this theme in "Systems Management and the Human Condition," in *Improving the Human Condition: Quality and Stability in Social Systems,* Richard F. Ericson, Ed., (Washington, DC: The George Washington University, Society for General Systems Research, 1979), Proceedings of the Silver Anniversary International Meeting, London, England, August 20-24, 1979.

Index